Table of Content

Chapter 1: Sheet metal Designing I..2

Chapter 2: Sheet metal Designing II..14

Chapter 3: Sheet metal Drafting..18

Chapter 4: Kinematics I..23

Chapter 5: Kinematics II...36

Chapter 6: Kinematics III..40

Chapter 7: Surfacing I...51

Chapter 8: Surfacing II..62

Chapter1: Sheet Metal Designing I

These are step-by-step instructions to create a sheet metal bracket in Catia V5. In this lesson, you will become familiar with some basic functions in the Generative Sheetmetal Design workbench. The following toolbars will be used.

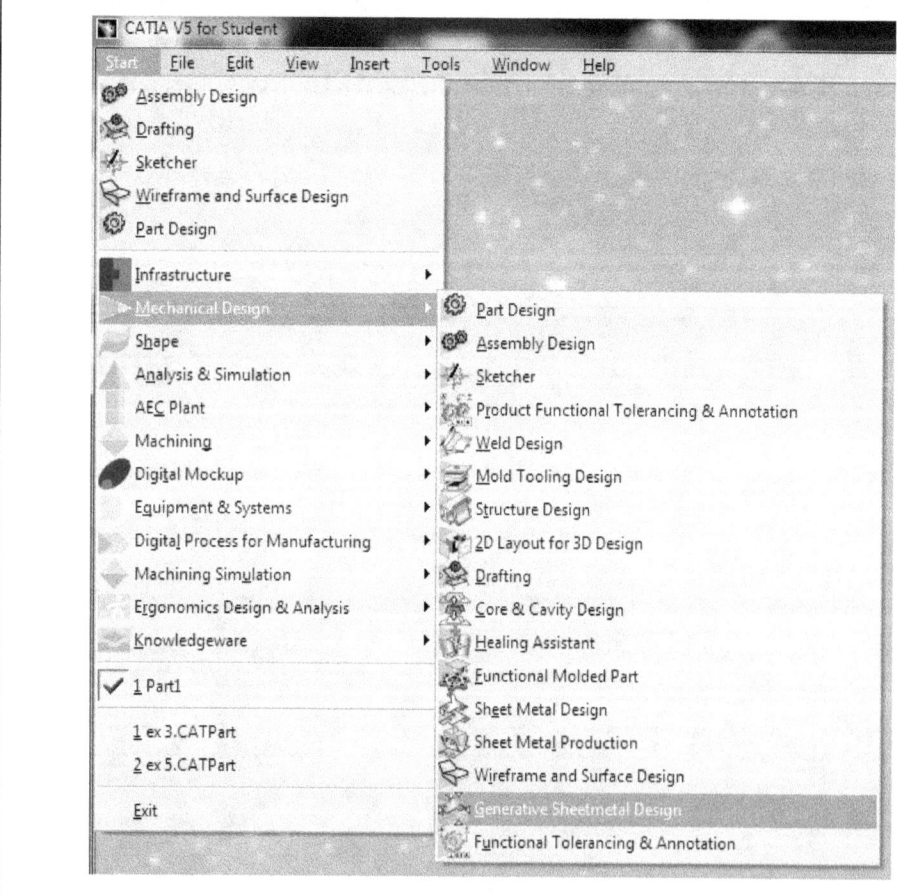

1. Select Generative Sheetmetal Design Workbench as shown.

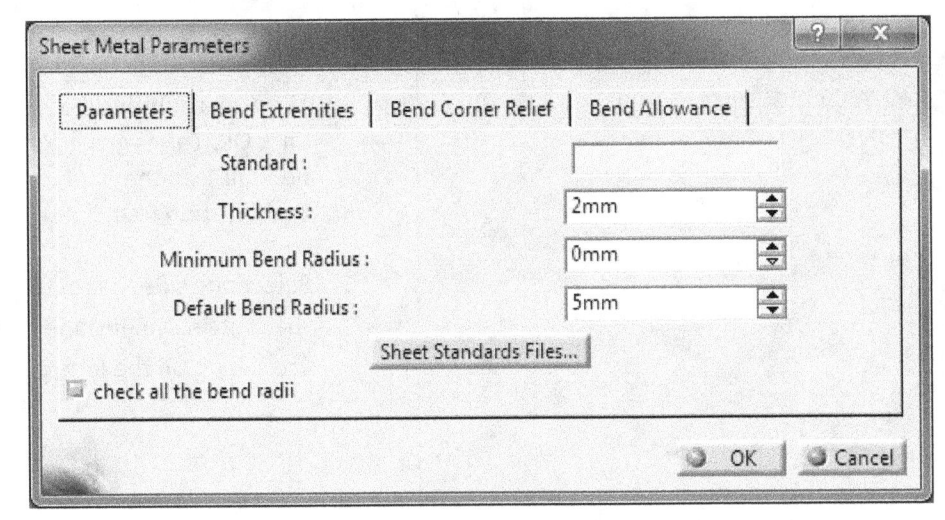

2. Select this icon (Sheet Metal Parameters) to set a parameter as shown.

3. Select the XY plane and create a sketch. Exit the sketcher when finished.

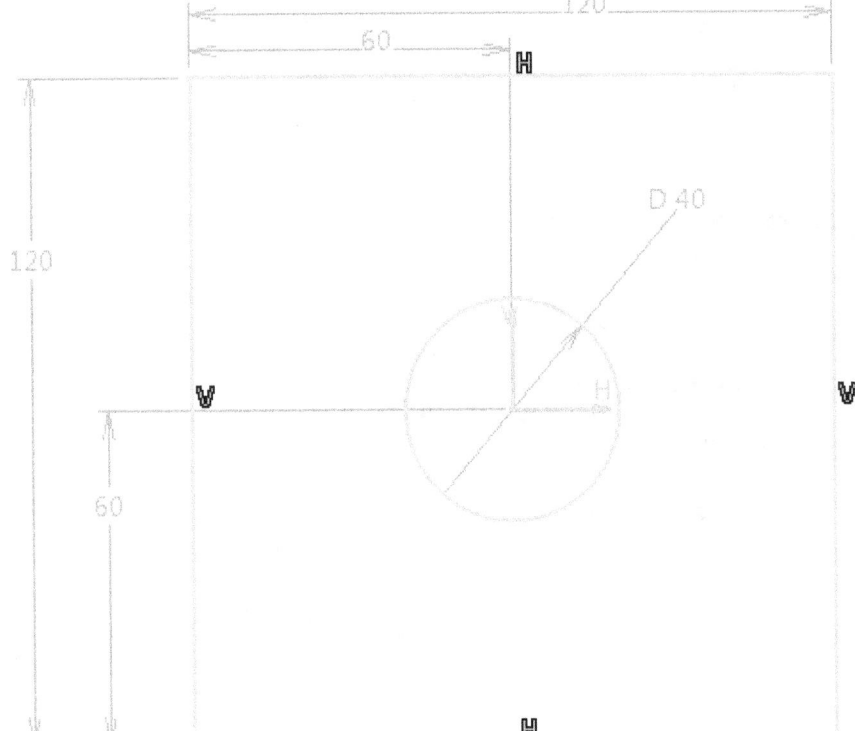

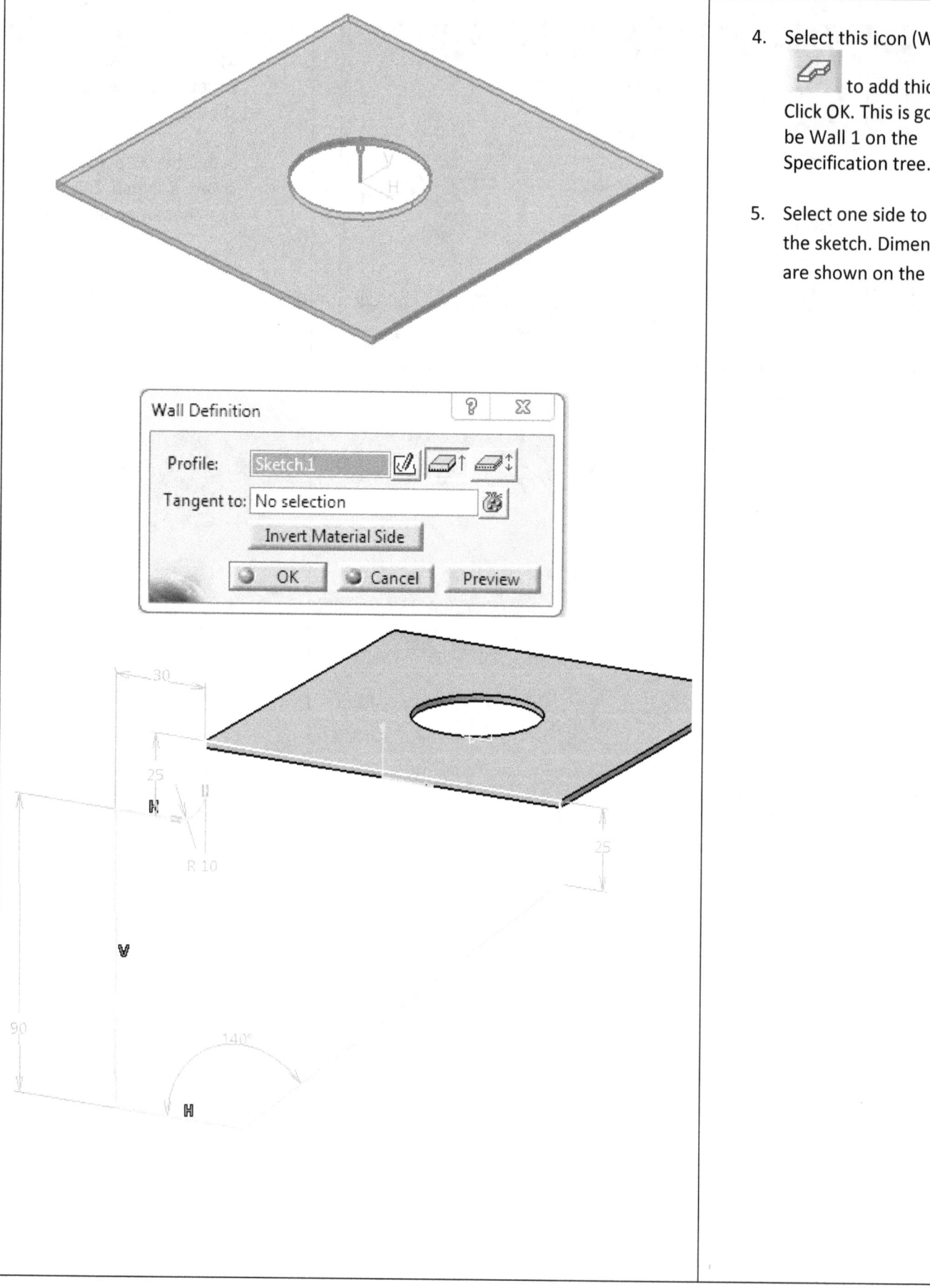

4. Select this icon (Wall) to add thickness. Click OK. This is going to be Wall 1 on the Specification tree.

5. Select one side to create the sketch. Dimensions are shown on the left.

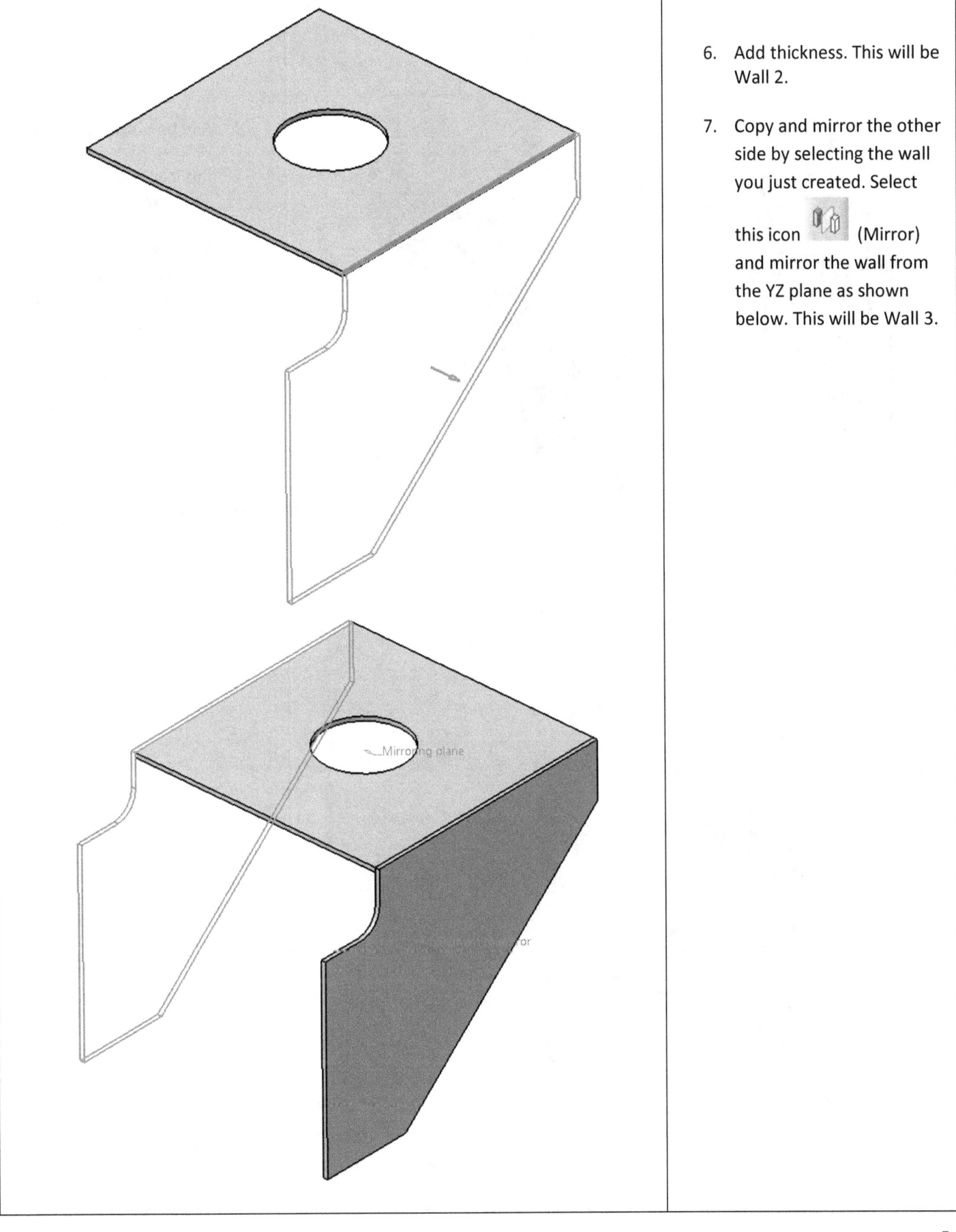

6. Add thickness. This will be Wall 2.

7. Copy and mirror the other side by selecting the wall you just created. Select this icon (Mirror) and mirror the wall from the YZ plane as shown below. This will be Wall 3.

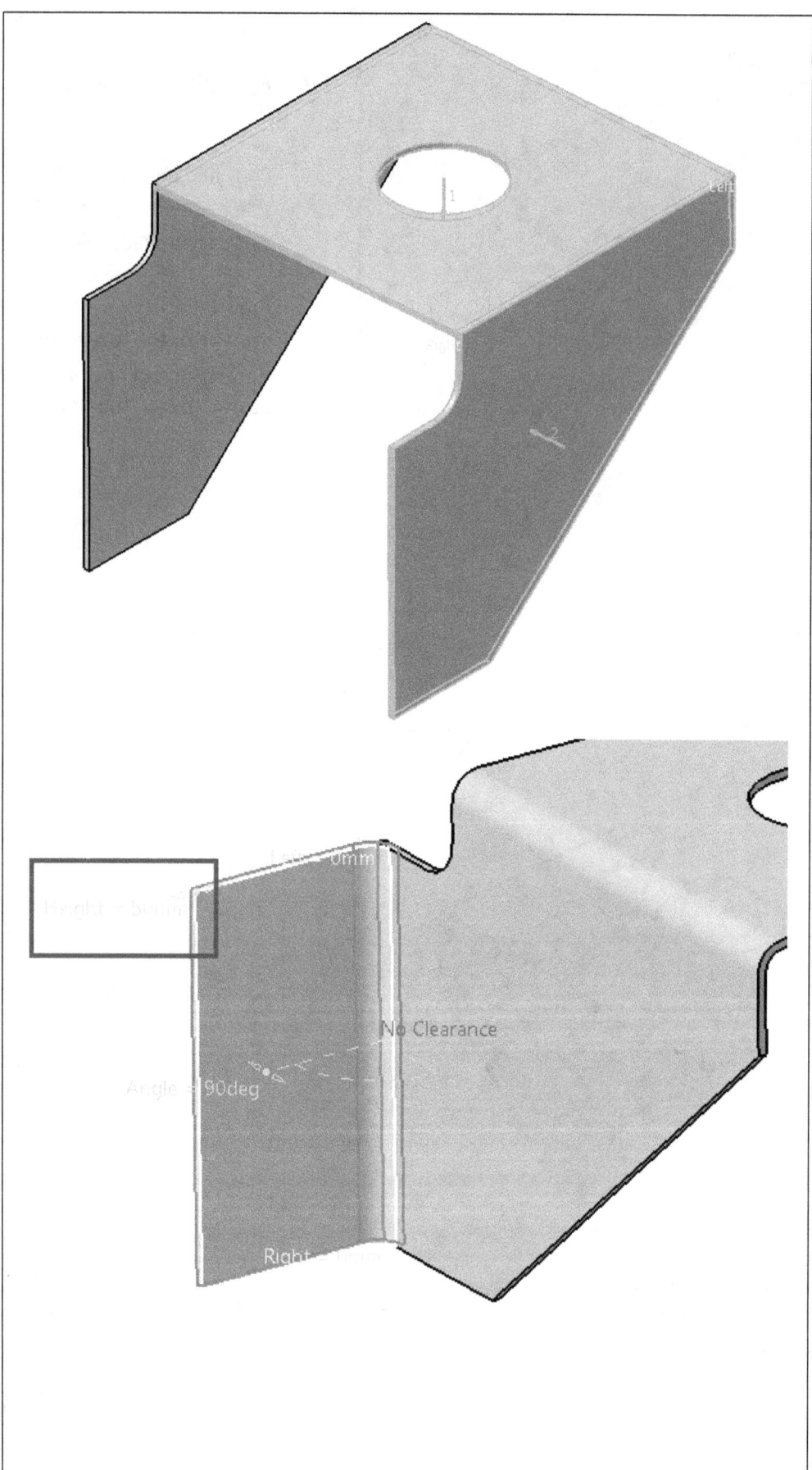

8. Add bends by selecting this icon (Bend) ⌐ Select Wall 1 for Support 1. Select Wall 2 for Support 2. Make sure the arrows are facing inward as shown. Repeat for the other side.

9. Add another wall by selecting this icon ◢ Select the edge of Wall 2 as shown.

10. Enter 50mm in Height as shown.

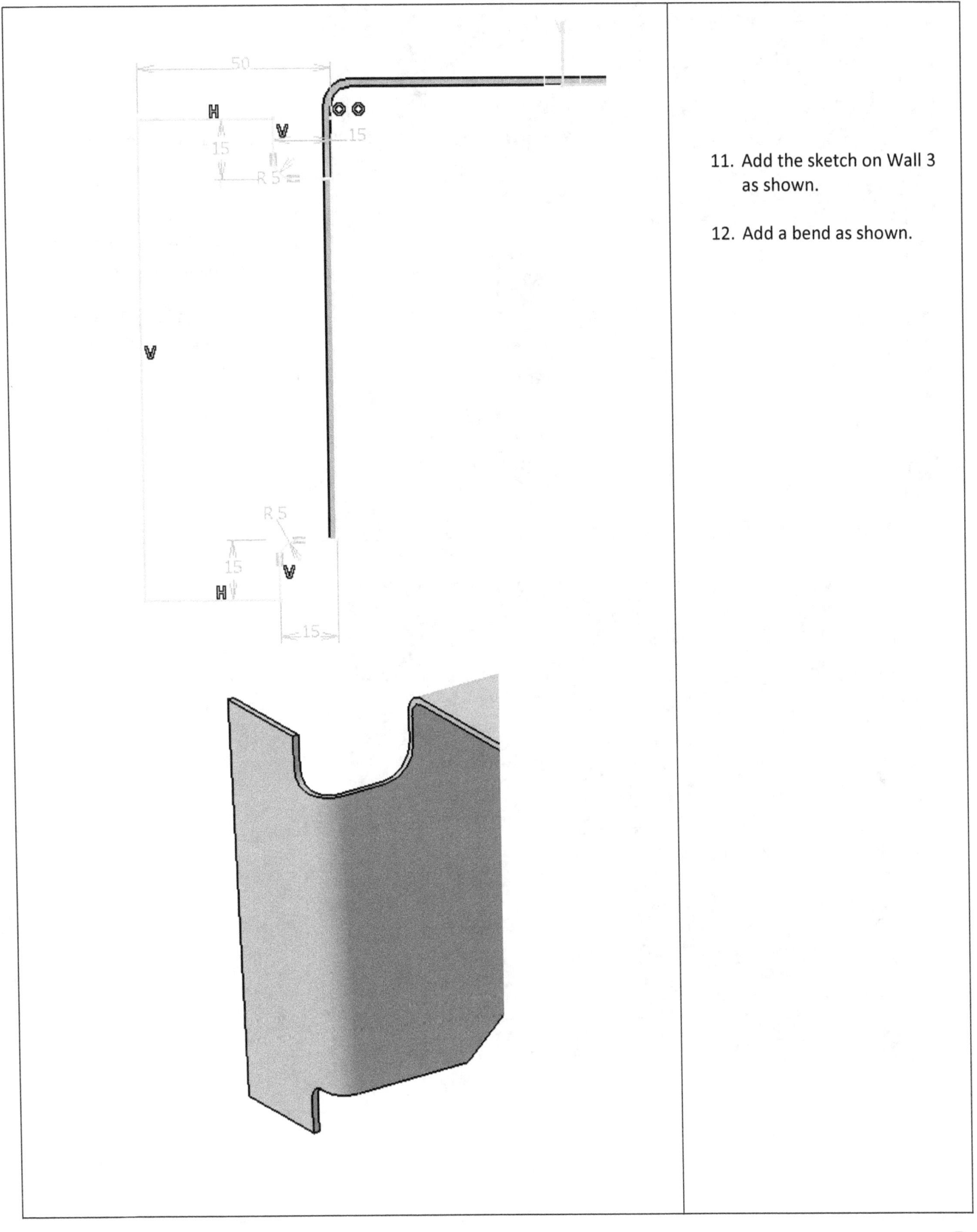

11. Add the sketch on Wall 3 as shown.

12. Add a bend as shown.

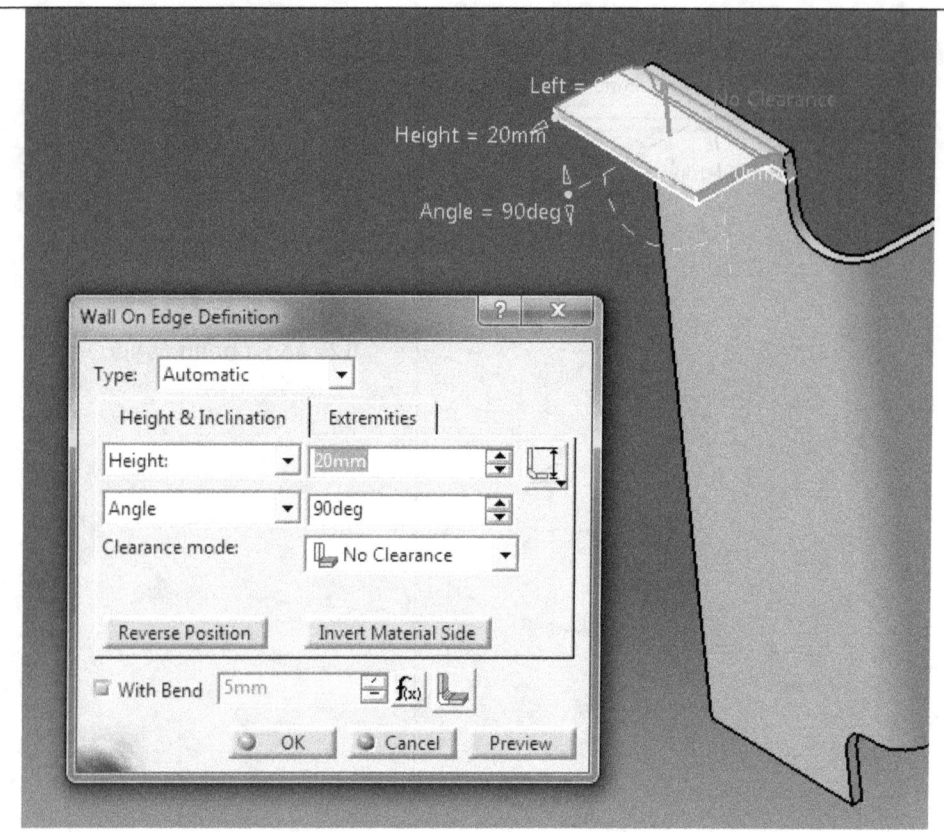

13. Add a flange by selecting this icon 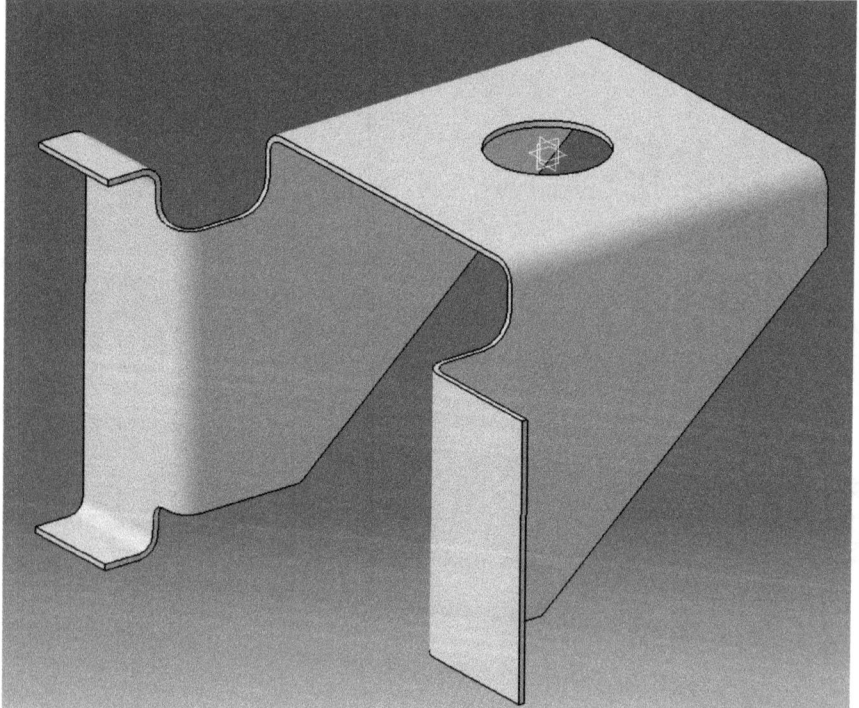 for strength as shown. This type of small wall id for structural reinforcement.

14. Repeat Step 13 for the bottom flange. It should be similar to the example on the left.

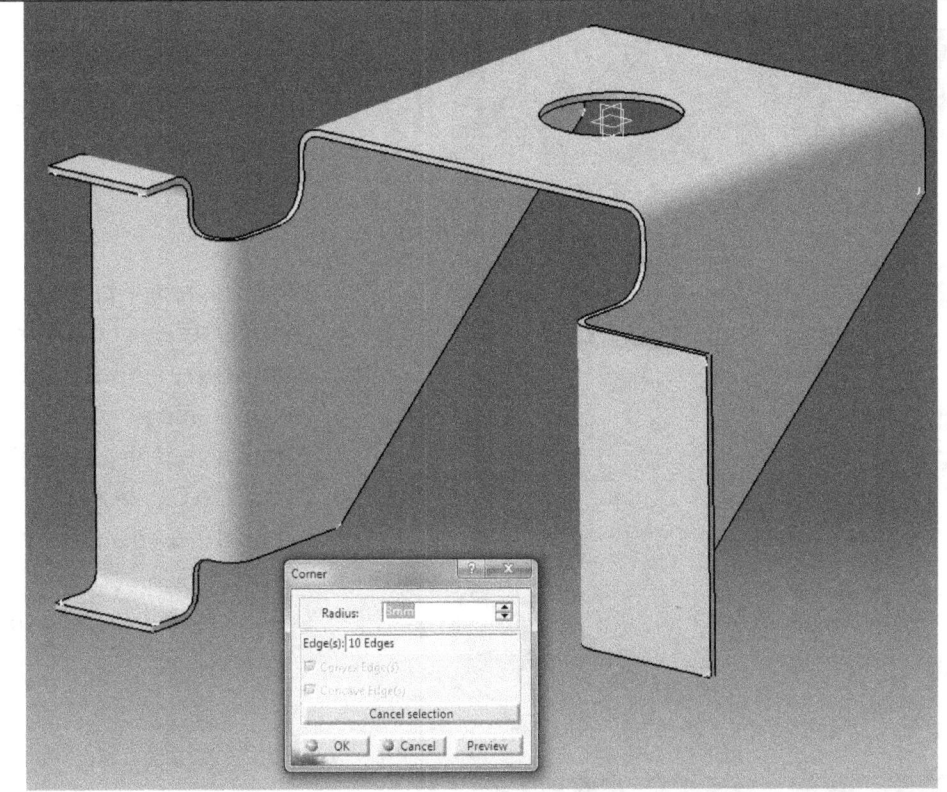

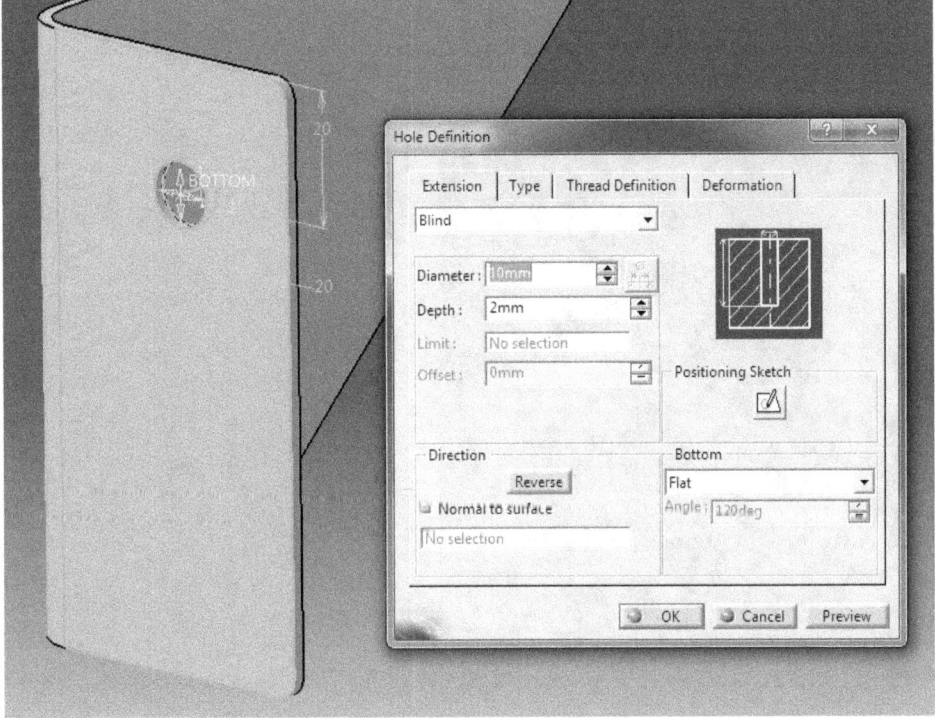

15. Add Corners by selecting this icon. Most brackets should not have sharp corners. Make each corner the same size. Let's add R3 on each corner.

16. Add holes. This can be done by either sketching or selecting this icon.

    Remember to add any features under Generative Sheet Metal Design Workbench, or when it is flattened, it won't work. Add two holes on the Wall on Edge as shown.

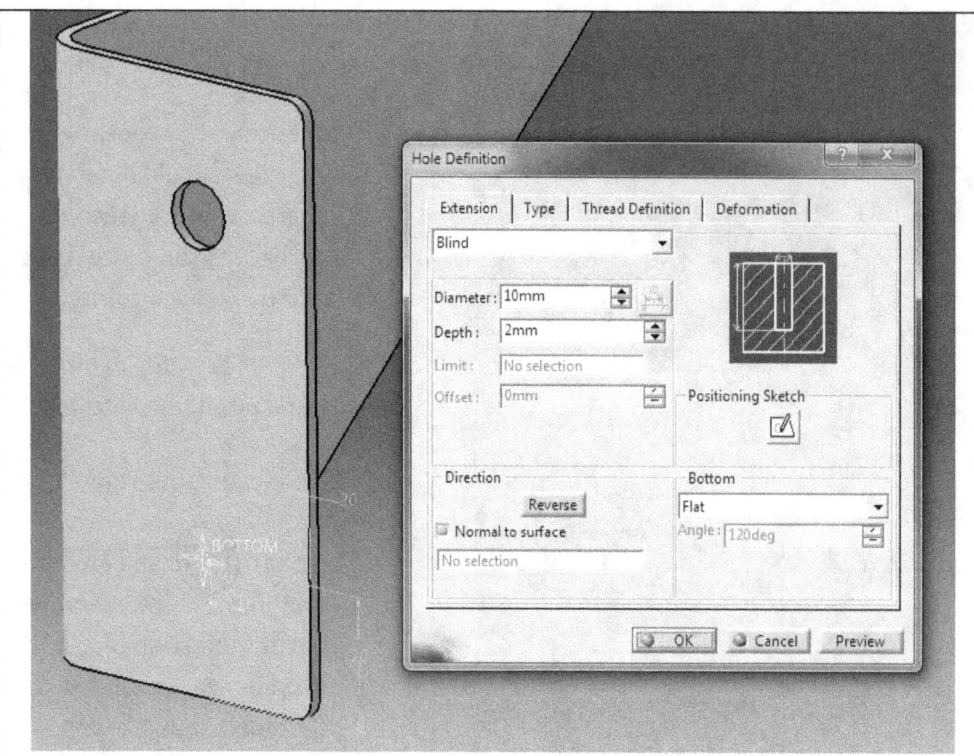

17. When creating a bracket, consider its weight. In the nearly every industry, light weight is important. Create an opening on Wall 2 and 3 so that less material is used and weight is lower. Create a sketch on Wall 2 similar to what's shown.

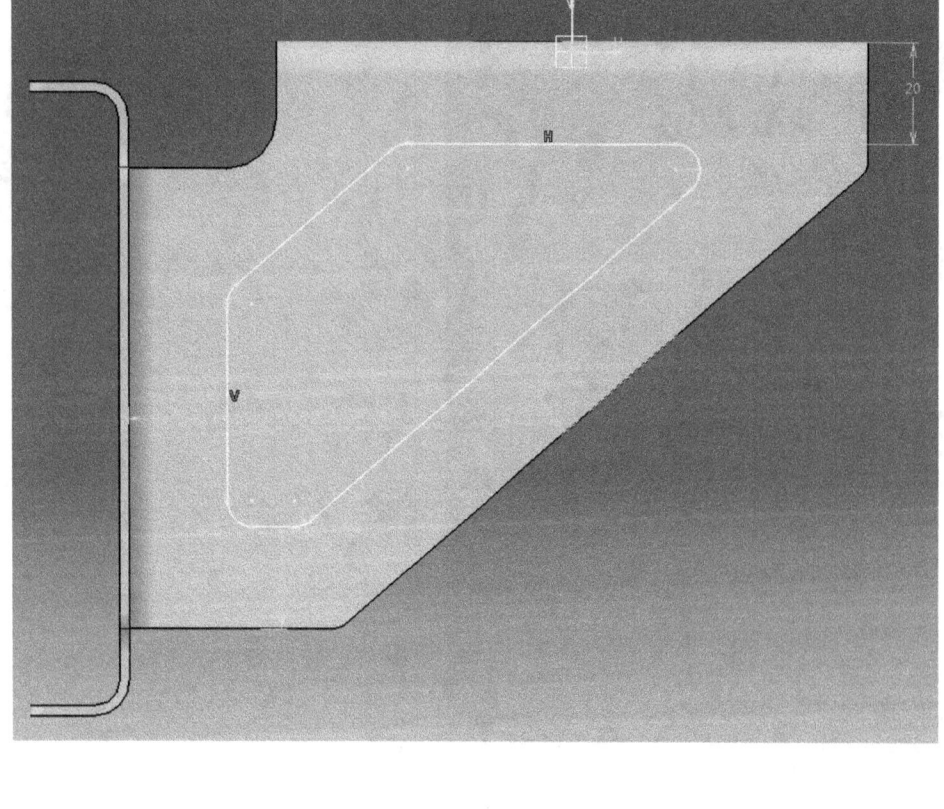

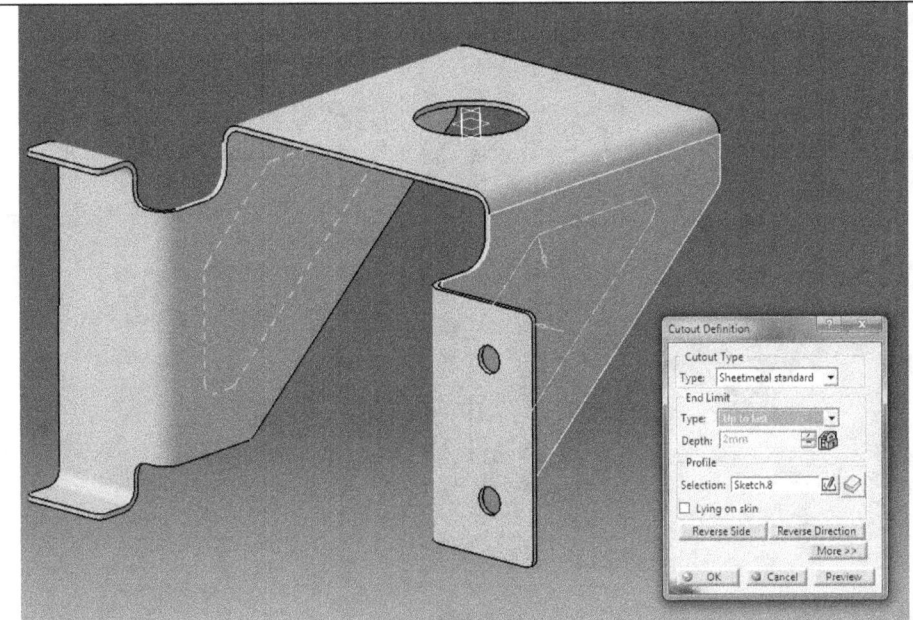

18. Select this icon and select "Up to last" to cut it all the way through.

Your final product should be similar to the example on the left.

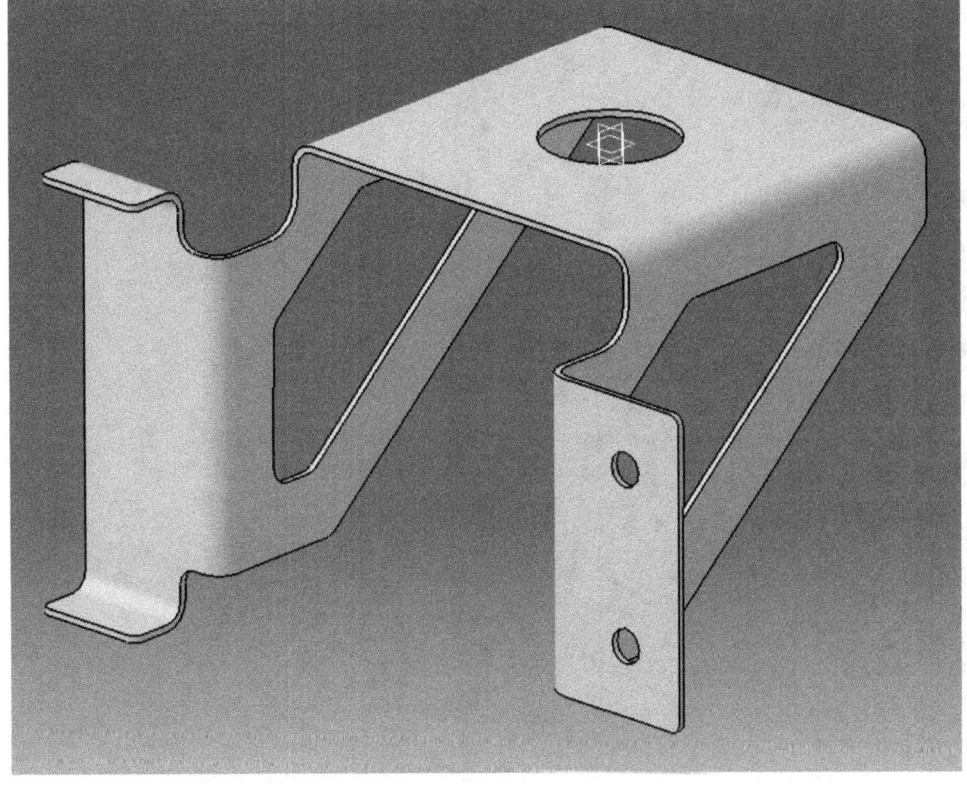

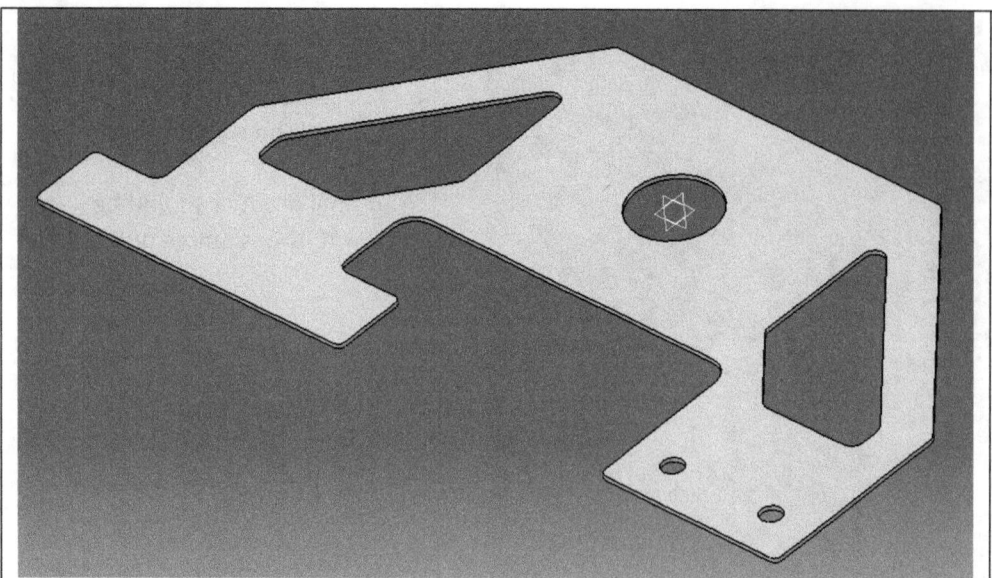

Flatten this part by clicking this icon ⚙ to see if the flat pattern does not cause interference.

You can click the same icon ⚙ to go back.

Save as "Sheet Metal 1."

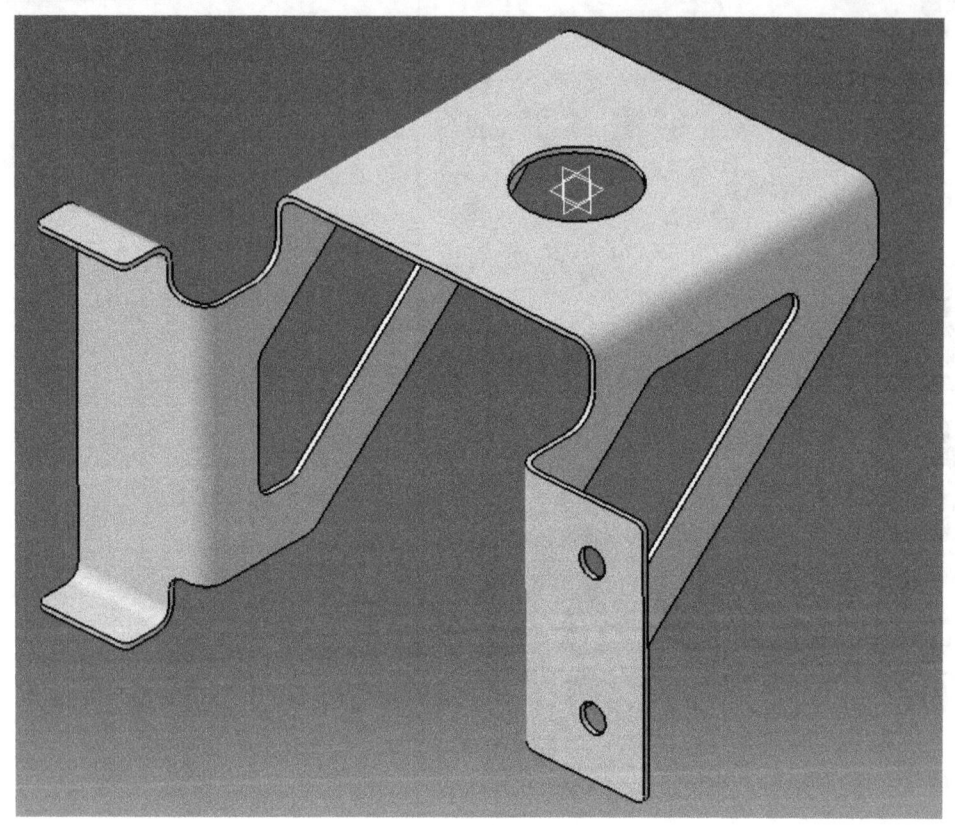

Chapter 1 Assignment

Based on the practice above, create the following bracket yourself. Dimensions are unimportant, so choose your own. Try your best to make a similar bracket to the one below. Have fun!

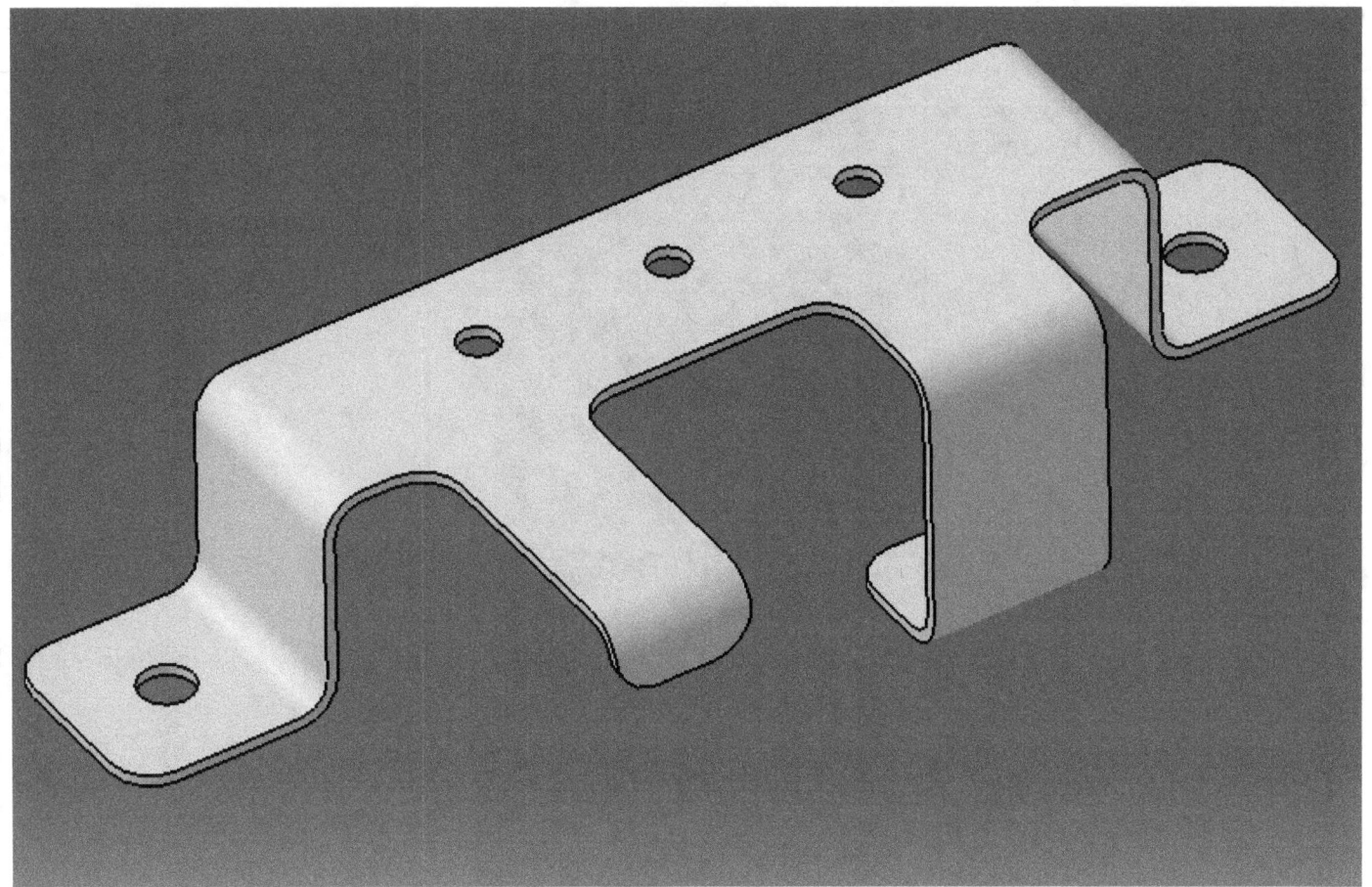

Chapter 2: Sheet metal Designing II

This is an introduction to some other functions that will be useful in the Generative Sheet Metal Design Workbench.

Use this icon (Extrusion) to extrude a pre-fixed shape. Example below.

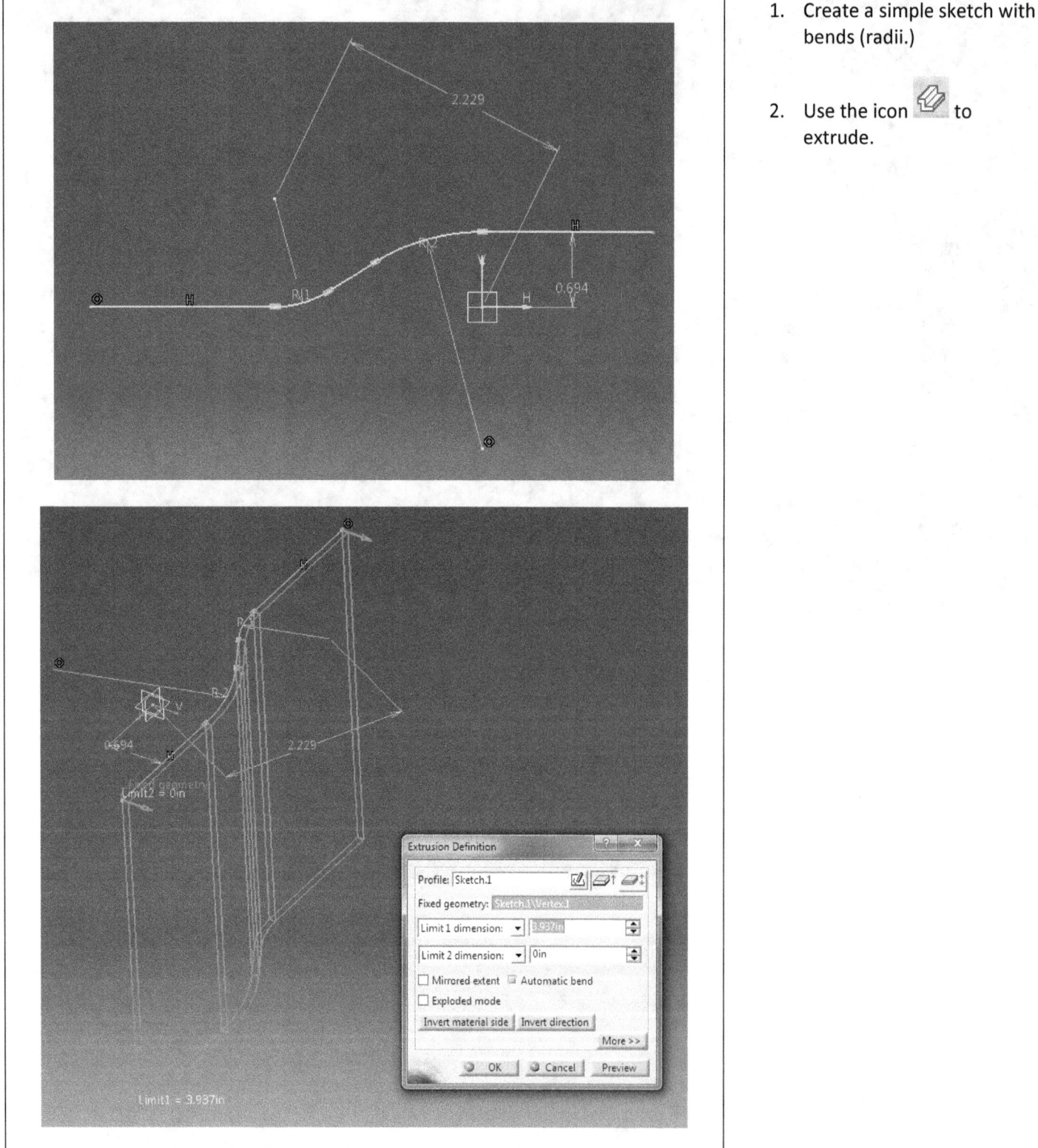

1. Create a simple sketch with bends (radii.)

2. Use the icon to extrude.

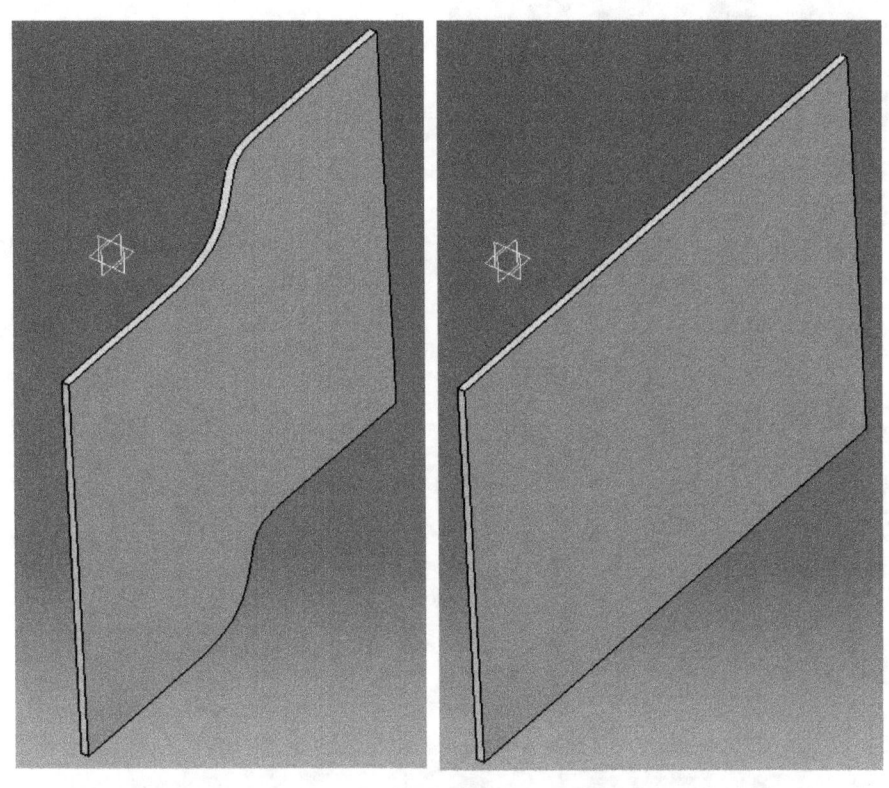

Use the Fold/Unfold icon to flatten it.

3. Use this icon (Bend from Flat) to add a new bend on the flat area. First, create a simple sketch (only a line) on any flat area.

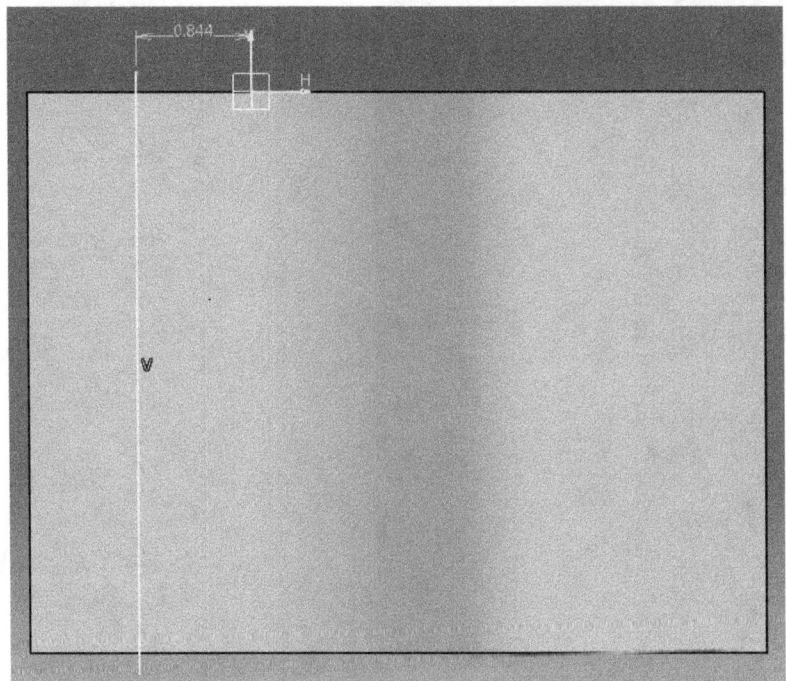

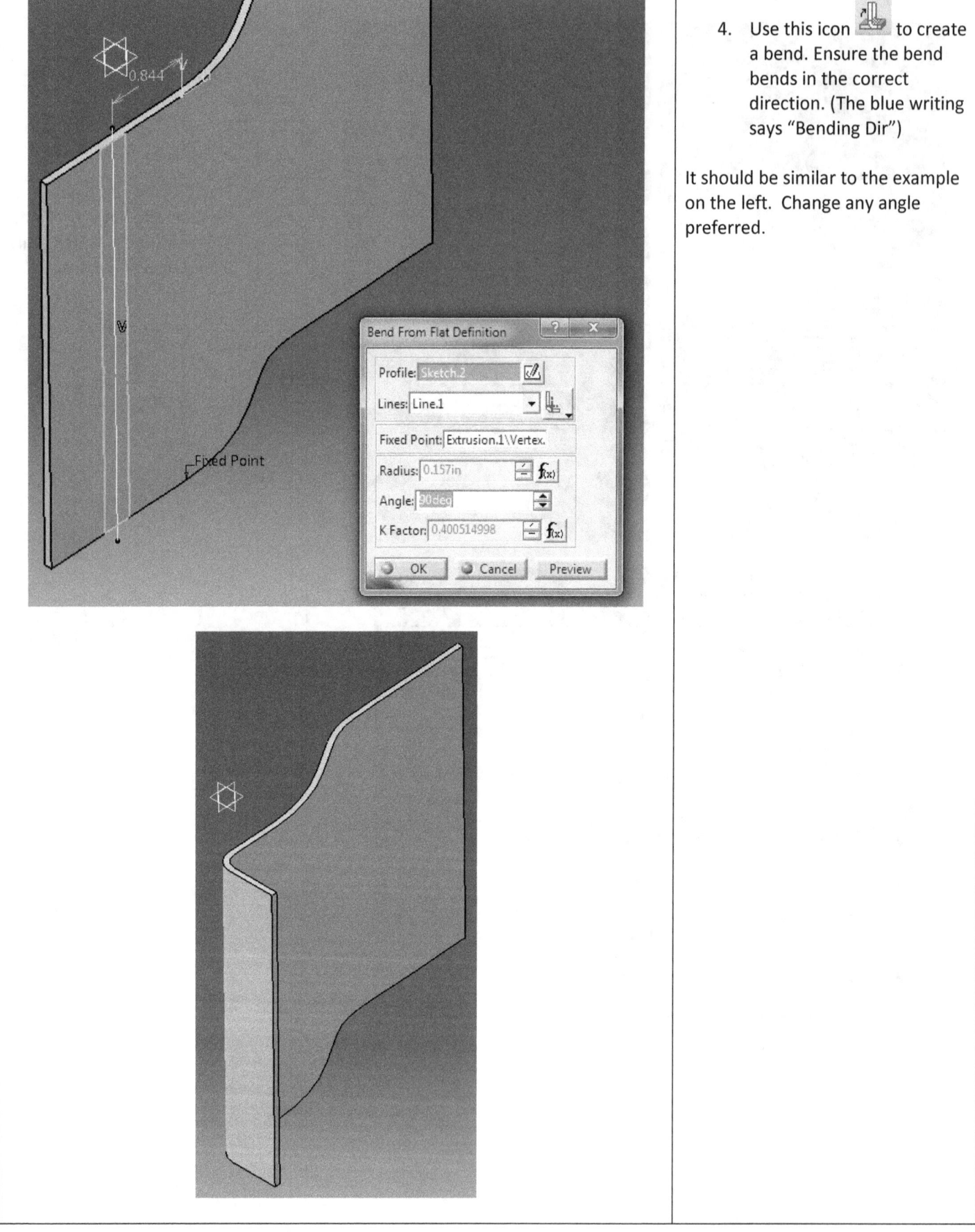

4. Use this icon to create a bend. Ensure the bend bends in the correct direction. (The blue writing says "Bending Dir")

It should be similar to the example on the left. Change any angle preferred.

Chapter 2 Assignment

As shown in class, make a bracket that connects those three objects. The design is up to you. Be creative and use less bends. In the industry, many bends and awkward bend angles are structurally unsound and thus are avoided.

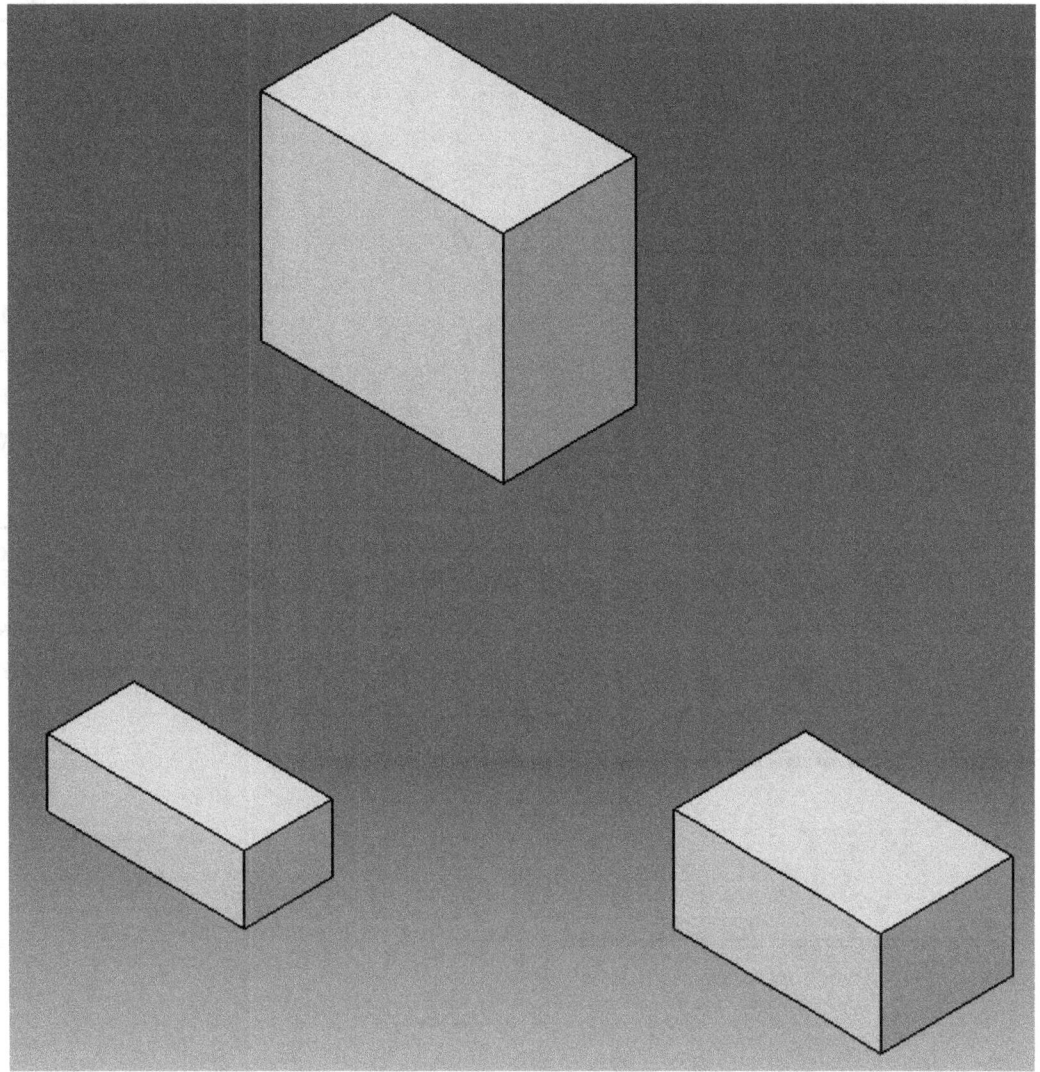

1. Go to Mechanical Design – Assembly Design

2. Highlight Product 1 on the Tree, and go to Insert – Existing component – Class drive – DGET 4470 and select "sheet metal assy.stp"

3. Highlight Product 1 on the Tree, and go to Insert – New Part and click OK.

4. Start your own design under the New Part.

Chapter 3: Sheet metal drafting

Inch and mm drawing sheets are available in the Class drive, DGET 4470. Use the proper sheets for completing this assignment.

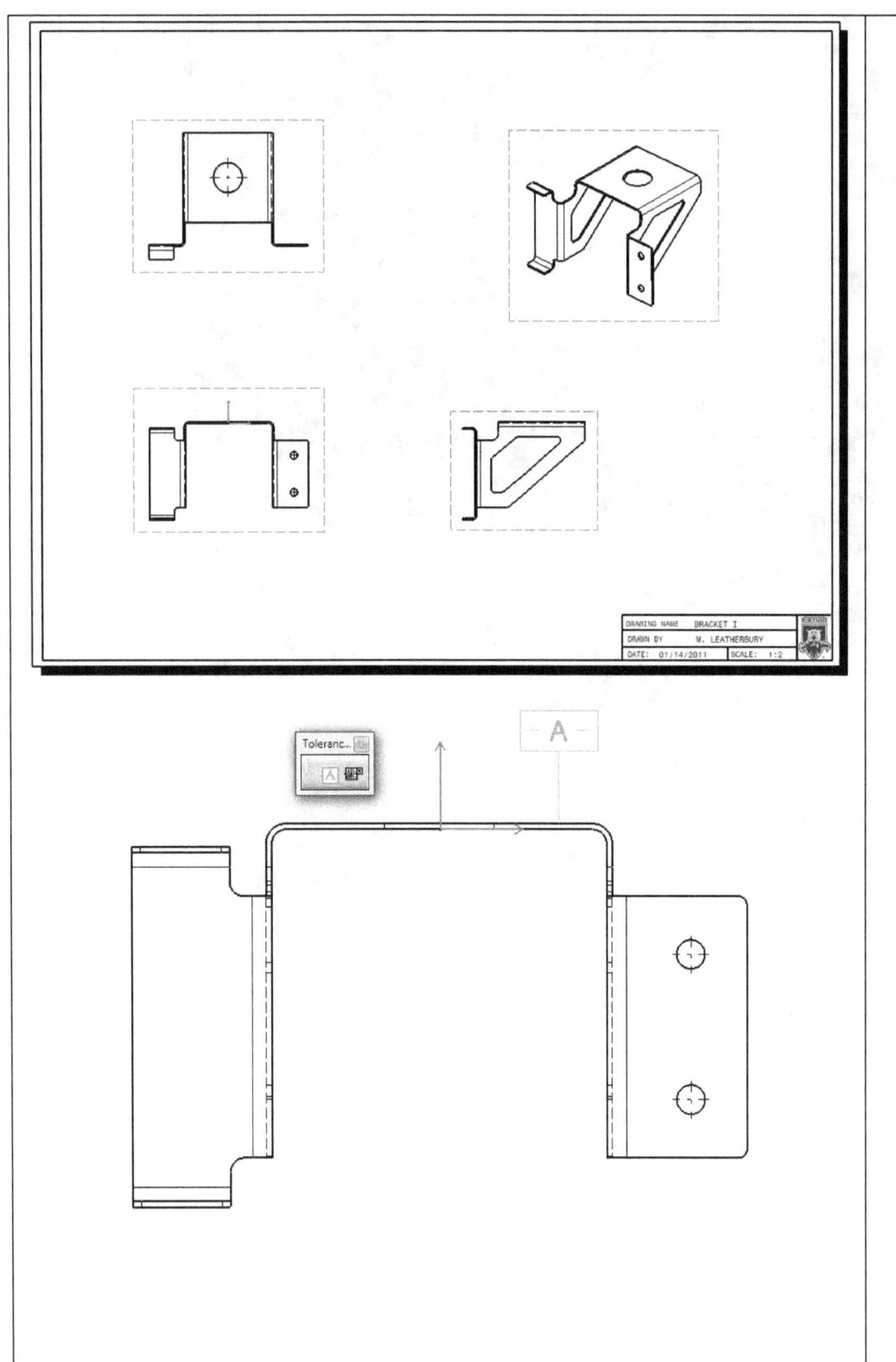

1. Create a drawing with all necessary views of Chapter 1, Sheet Metal 1. Complete dimensioning using the datum dimensioning method. Choose the appropriate datums and finish dimensioning accordingly. The thickness of the material should not be selected as a datum because it is too thin to be a datum.

2. Add datums. Example shown on the left.

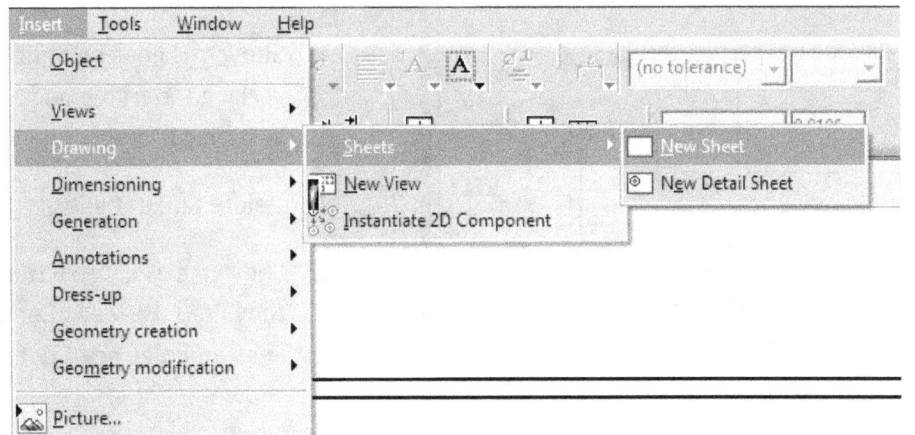

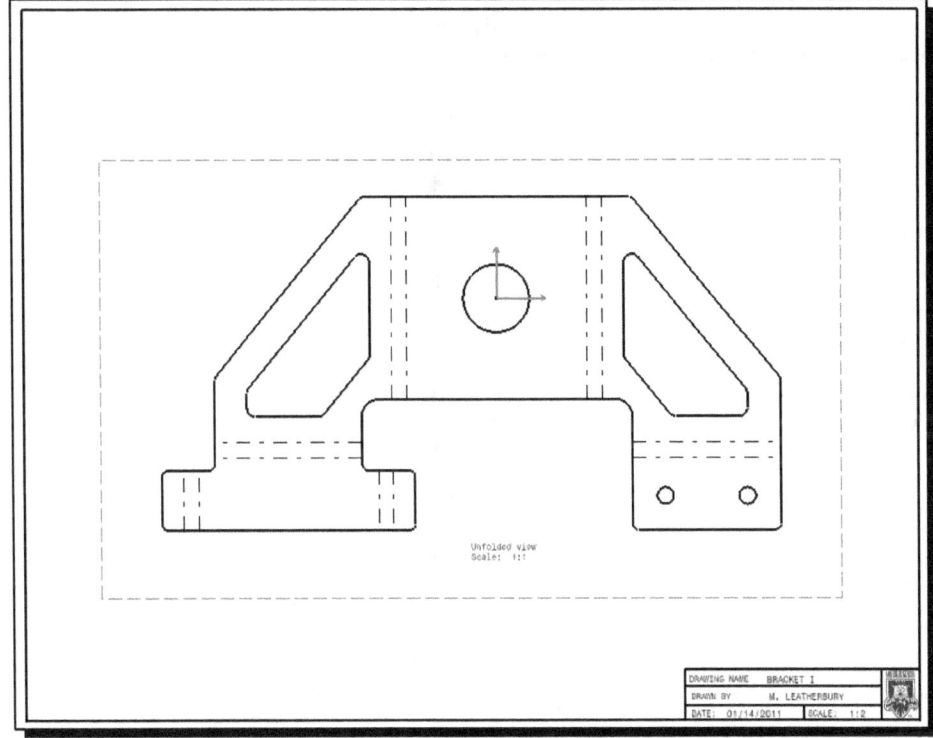

3. Complete the title block before adding another sheet so that all information will be copied. Everything should be written in upper case as shown below.

4. Insert another sheet. Notice the second page title block is already filled.

5. Add a flattened view on the new sheet. Select this icon and select 3D.

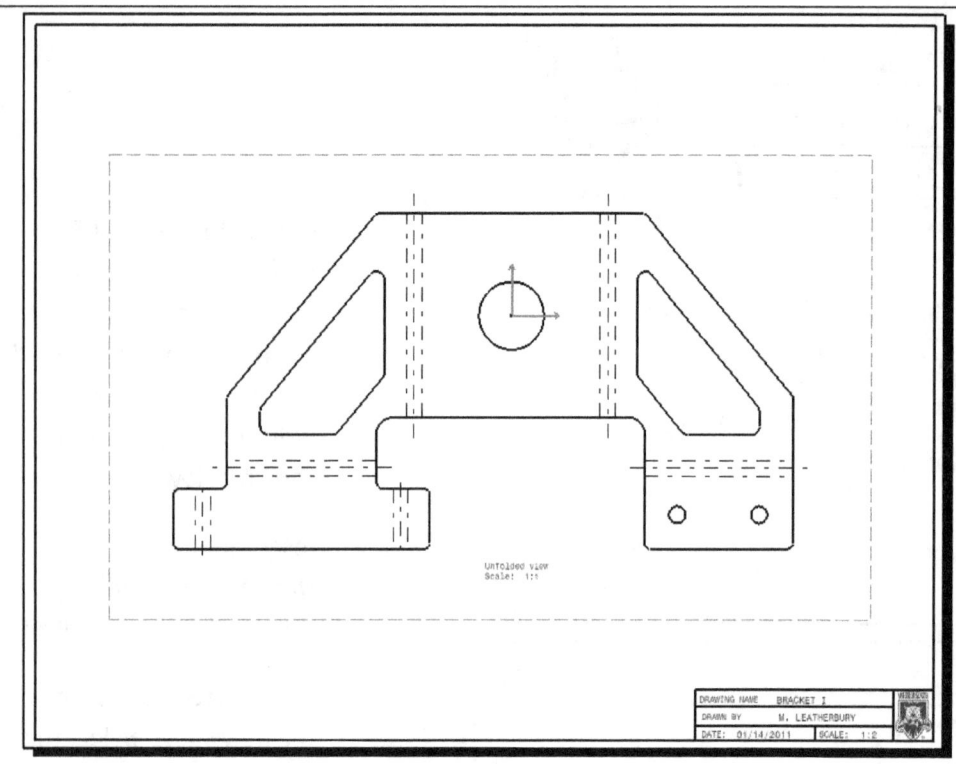

6. Show axis lines by right clicking the frame and selecting Properties.

7. Delete tangent lines.

8. Shorten axis lines so that they fit in the view. Center lines on holes need not be shown.

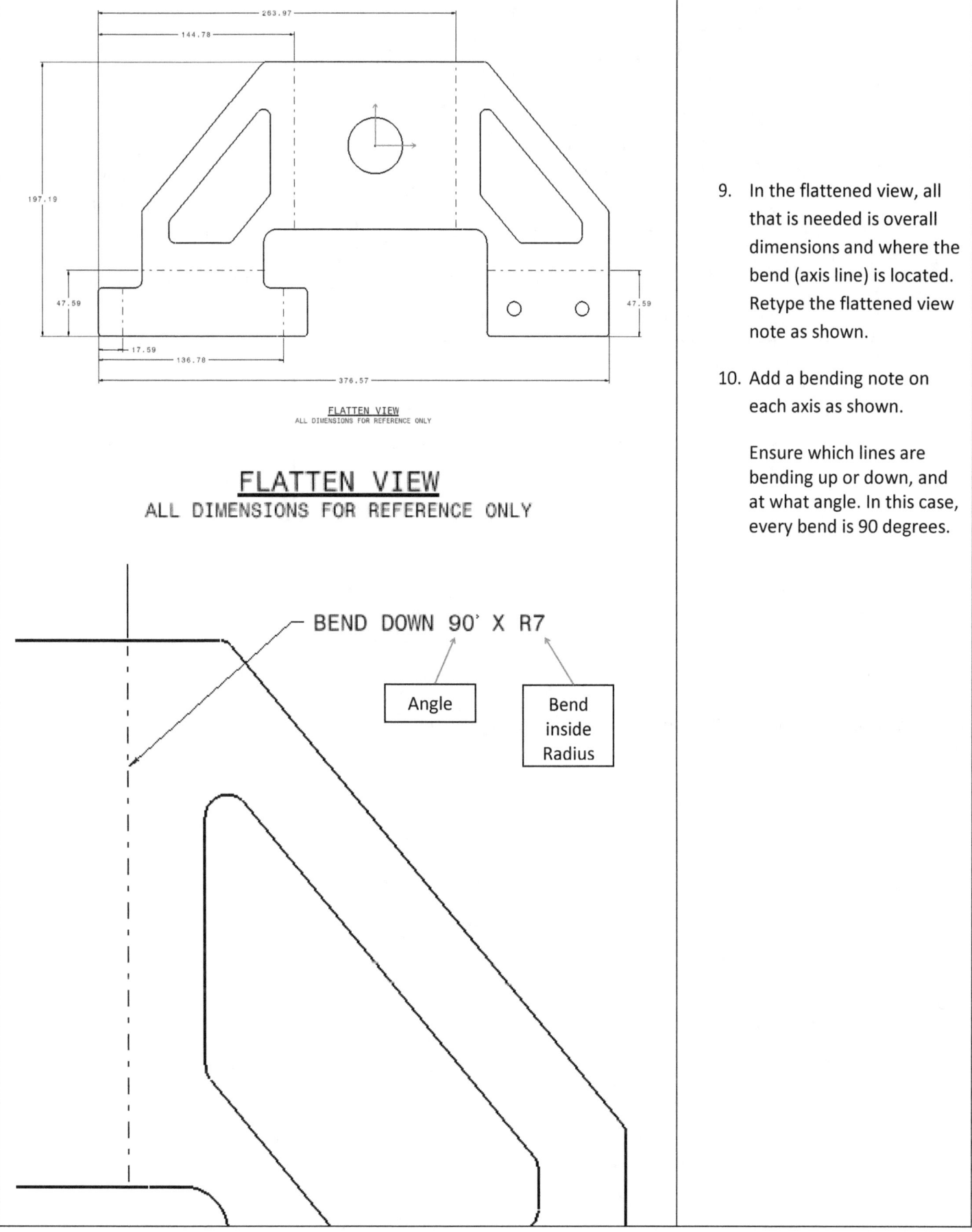

9. In the flattened view, all that is needed is overall dimensions and where the bend (axis line) is located. Retype the flattened view note as shown.

10. Add a bending note on each axis as shown.

   Ensure which lines are bending up or down, and at what angle. In this case, every bend is 90 degrees.

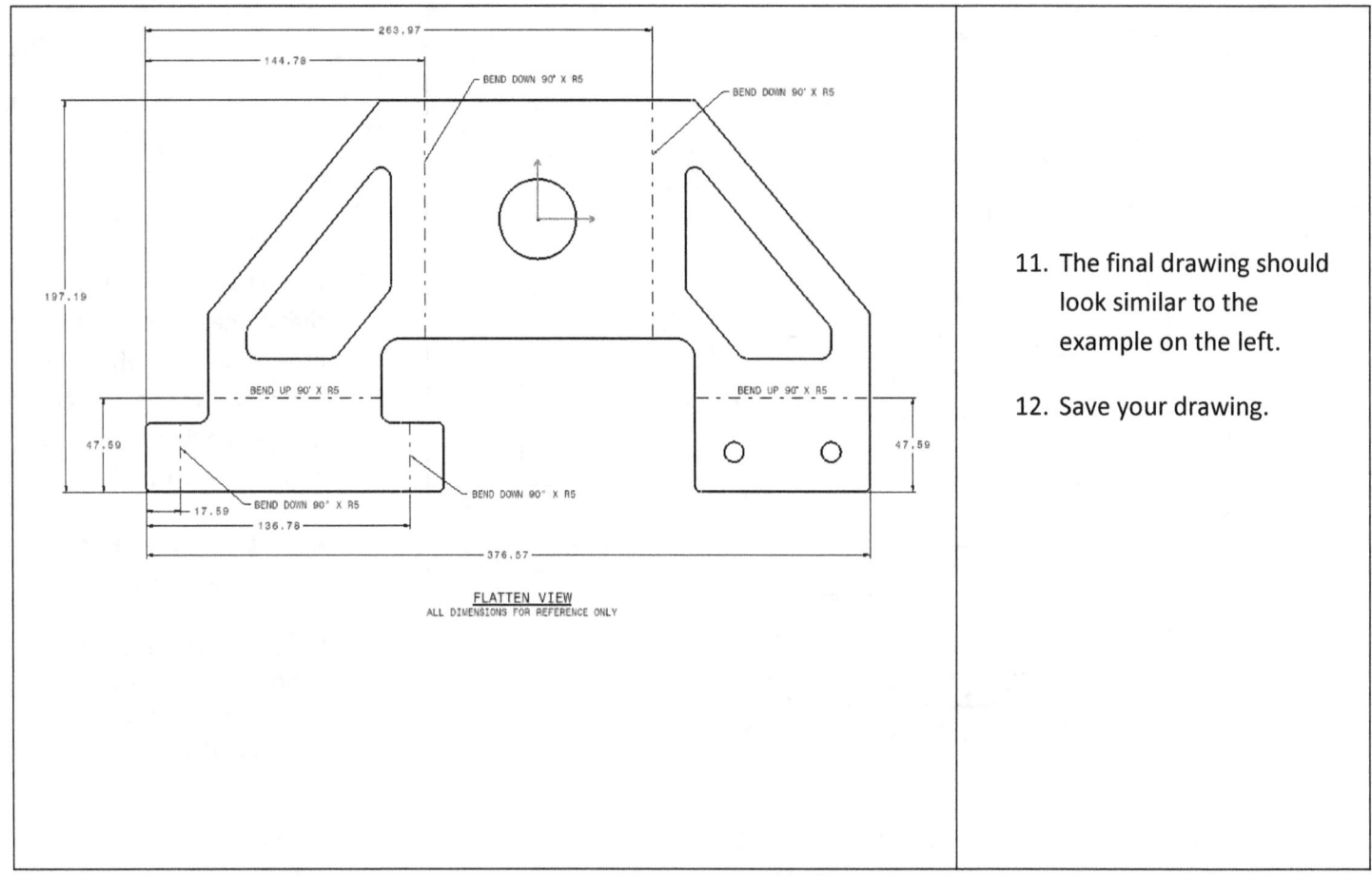

11. The final drawing should look similar to the example on the left.

12. Save your drawing.

Chapter 3 Assignment :

Create a drawing for each bracket you made in Chapter 2 assignment.

Chapter 4: Kinematics I

Prismatic Joint

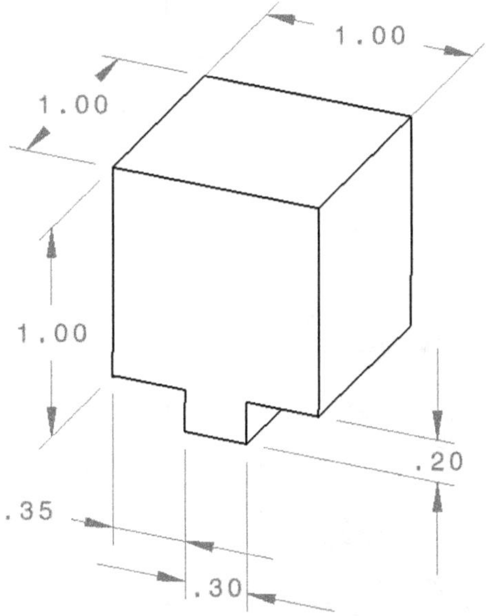

Key

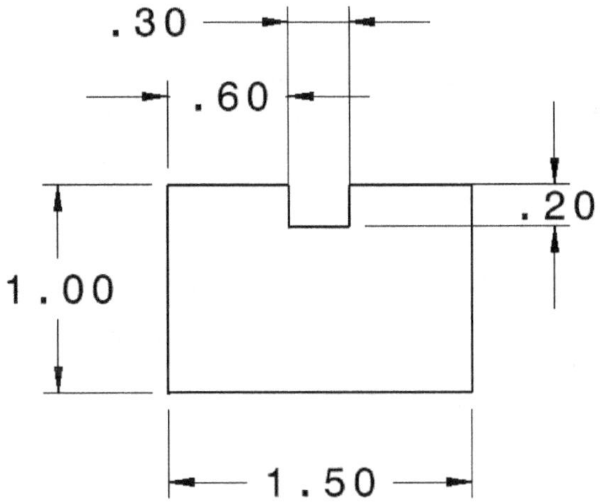

Base: Length 10"

Create two parts as shown in inches.

Create an assembly of these parts.

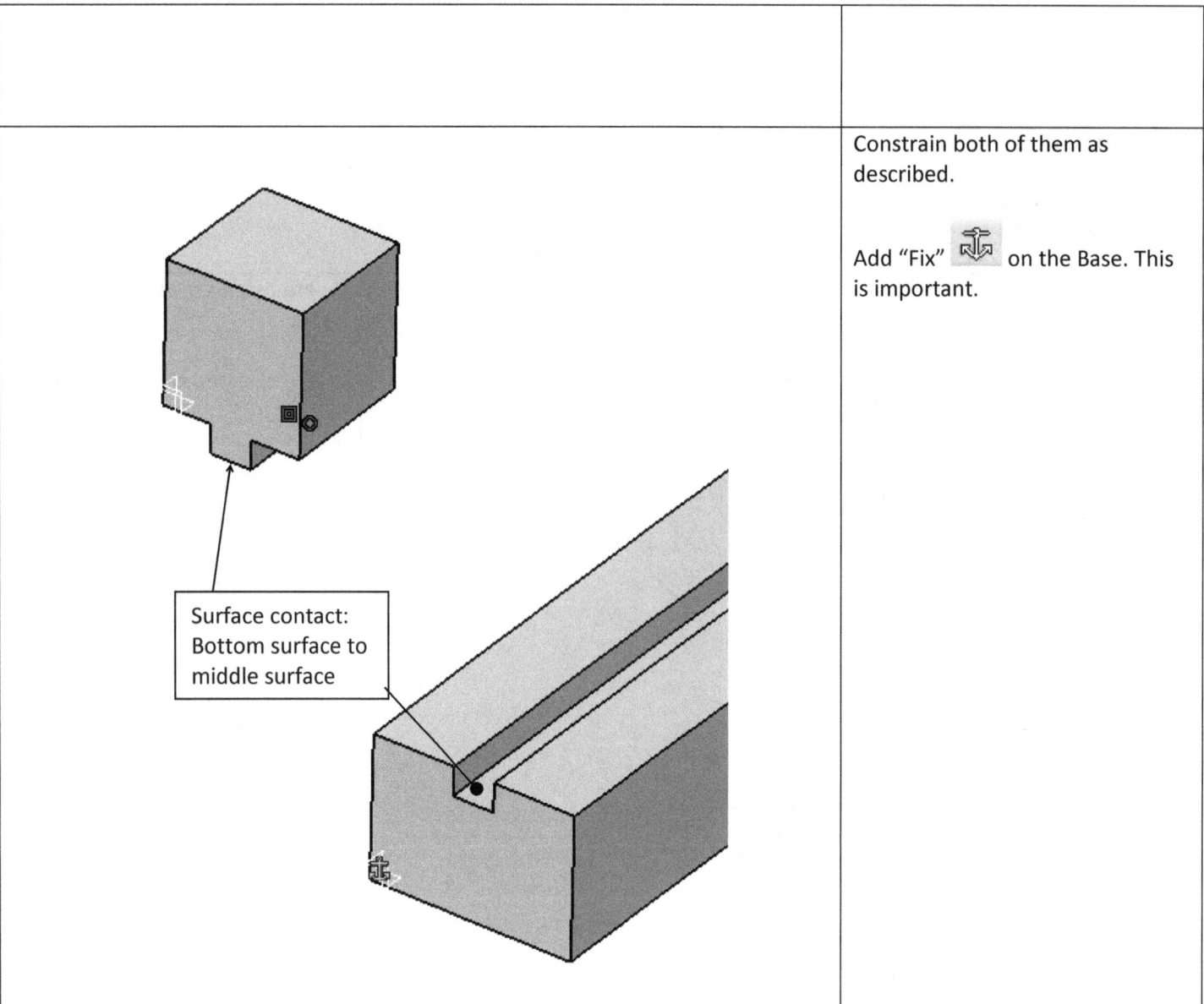

Surface contact: Bottom surface to middle surface

Constrain both of them as described.

Add "Fix" on the Base. This is important.

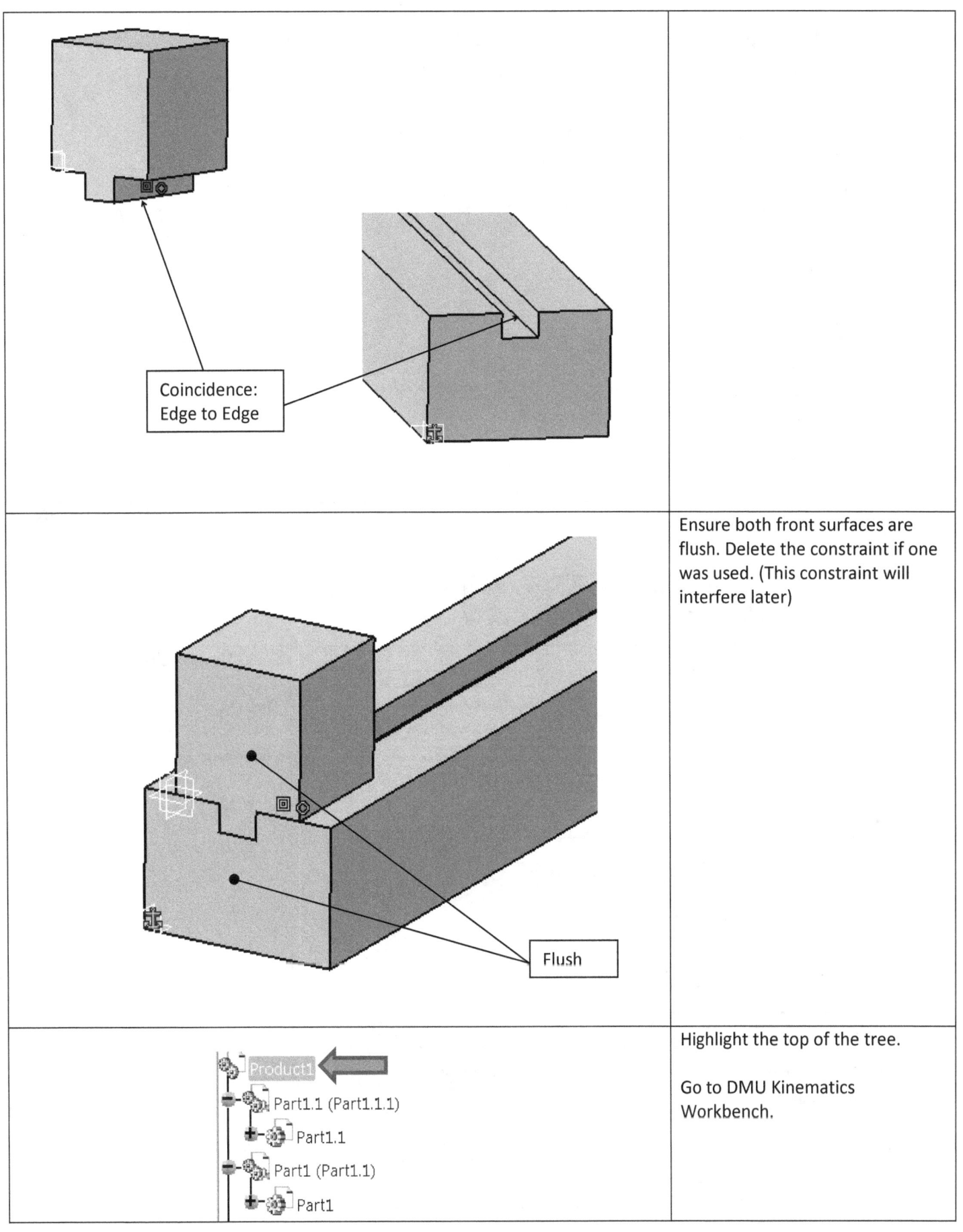

Ensure both front surfaces are flush. Delete the constraint if one was used. (This constraint will interfere later)

Highlight the top of the tree.

Go to DMU Kinematics Workbench.

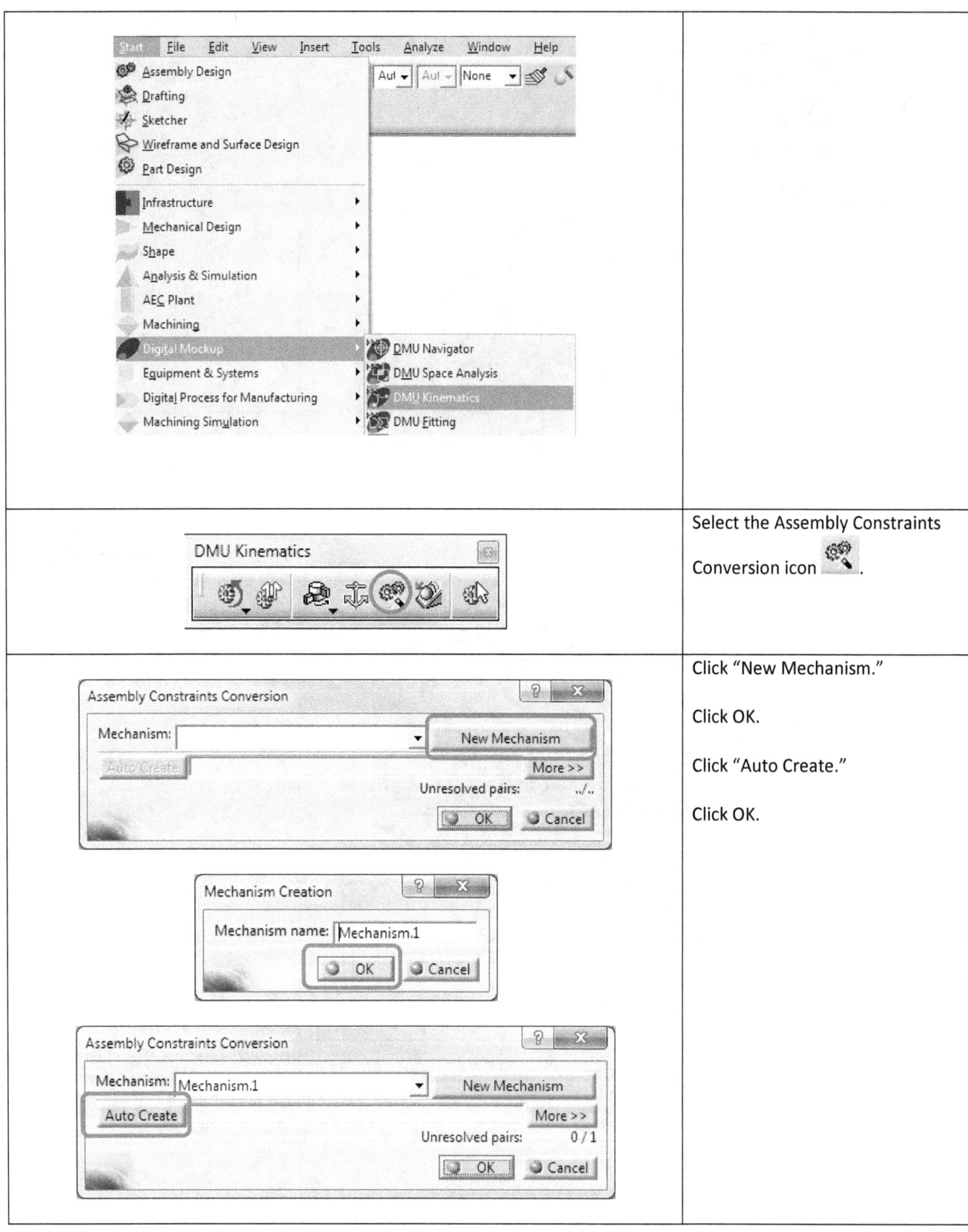

Select the Assembly Constraints Conversion icon.

Click "New Mechanism."

Click OK.

Click "Auto Create."

Click OK.

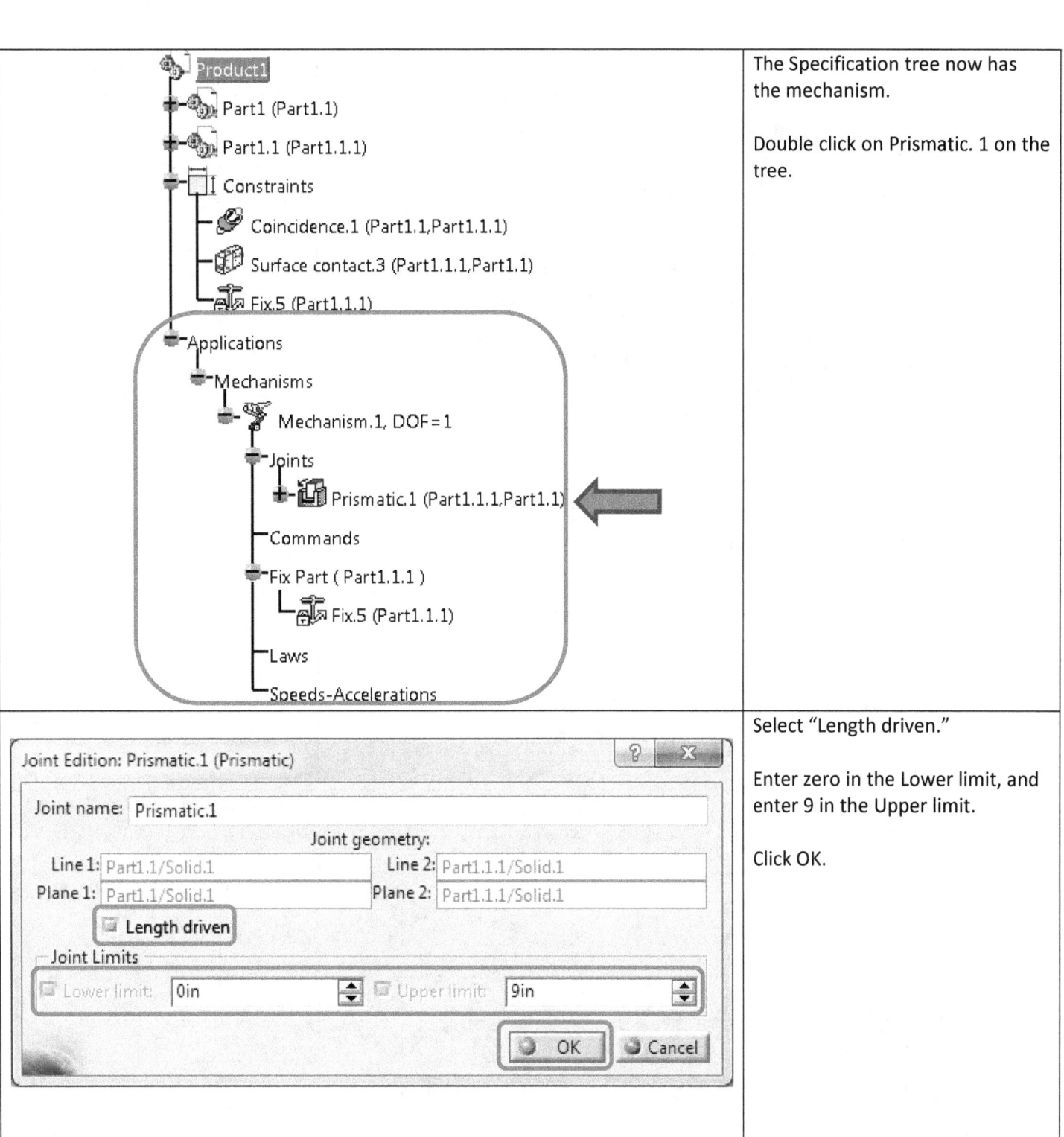

The Specification tree now has the mechanism.

Double click on Prismatic.1 on the tree.

Select "Length driven."

Enter zero in the Lower limit, and enter 9 in the Upper limit.

Click OK.

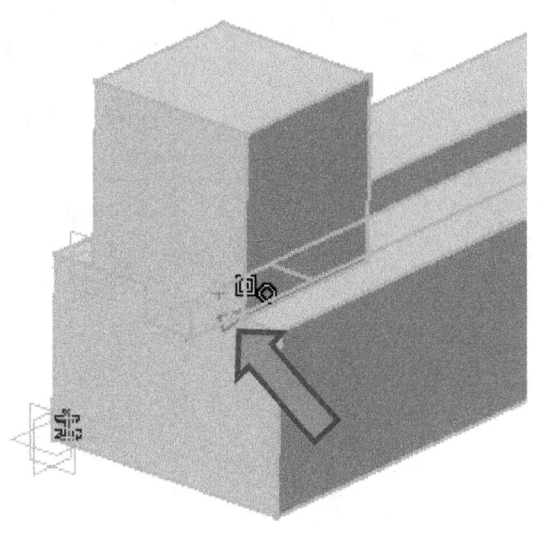

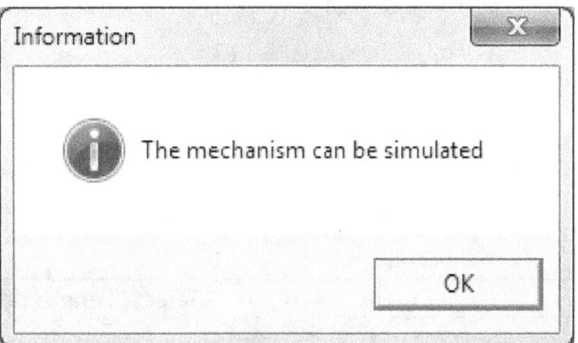

Ensure that blue arrow is facing the right direction as shown.

Now the mechanism can be simulated.

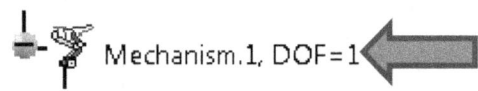

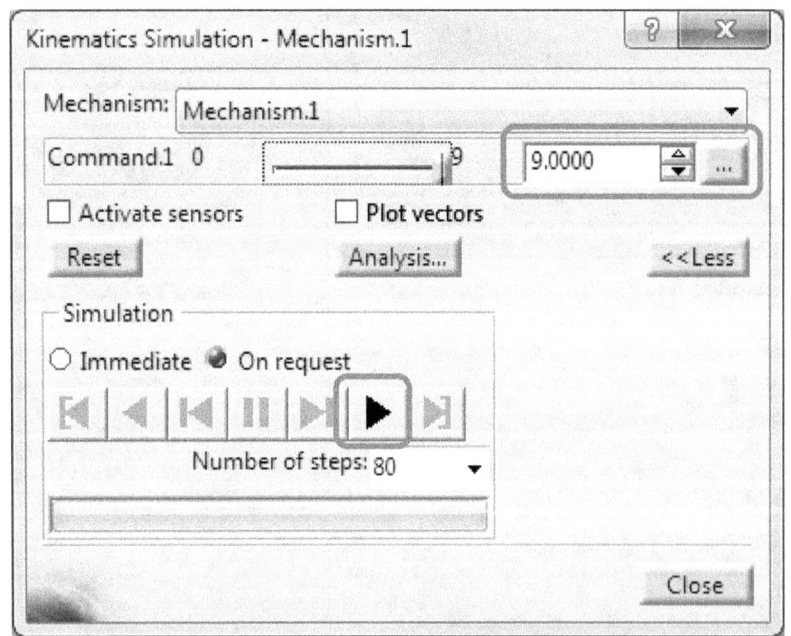

Double click Mechanism 1 on the tree.

Enter 9 in the Command box.

Click Play.

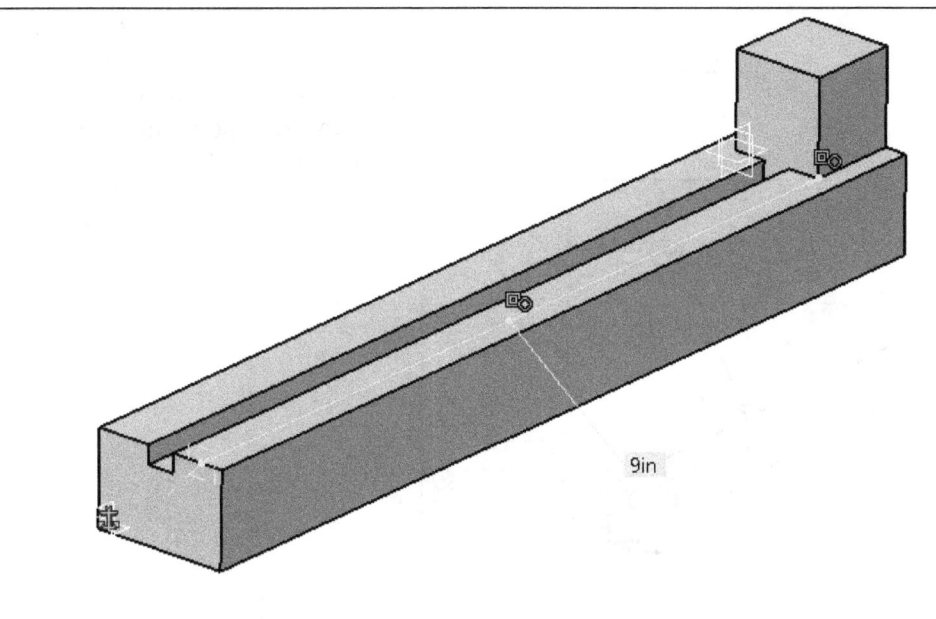

The Key should have moved 9 inches from the starting point.

"Number of steps" will make it go faster or slower. The larger the number, the slower the simulation.

Enter different values to see how it moves.

Save the file.

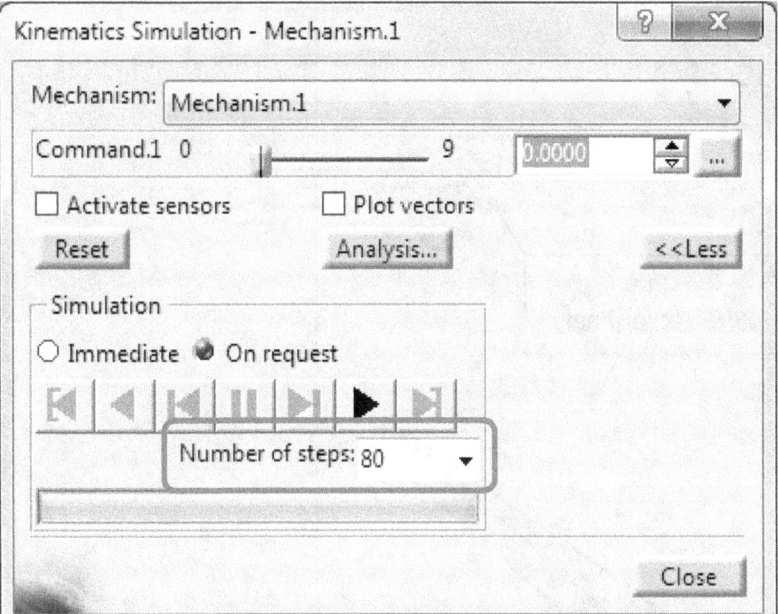

Revolute joint

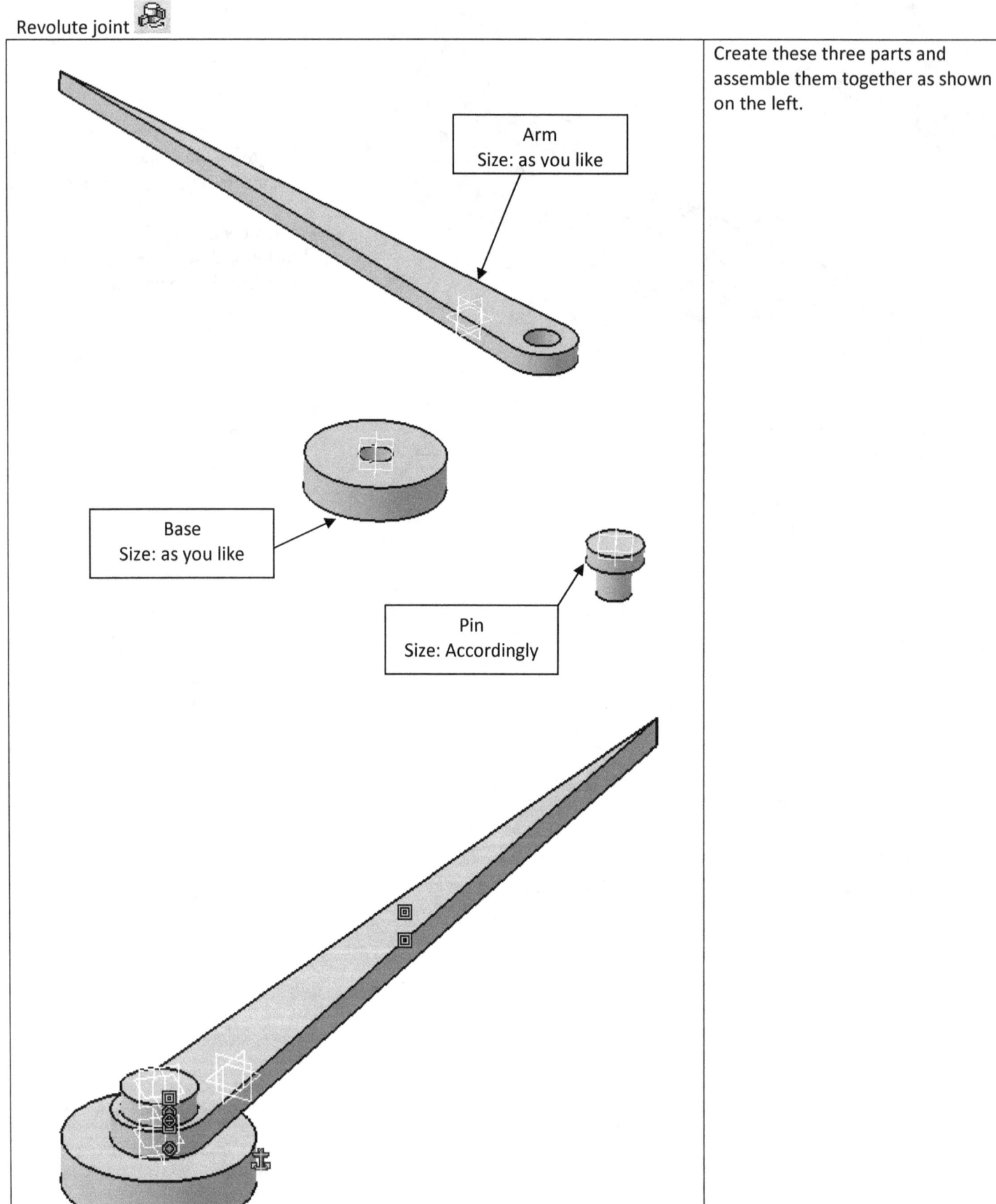

Create these three parts and assemble them together as shown on the left.

Arm
Size: as you like

Base
Size: as you like

Pin
Size: Accordingly

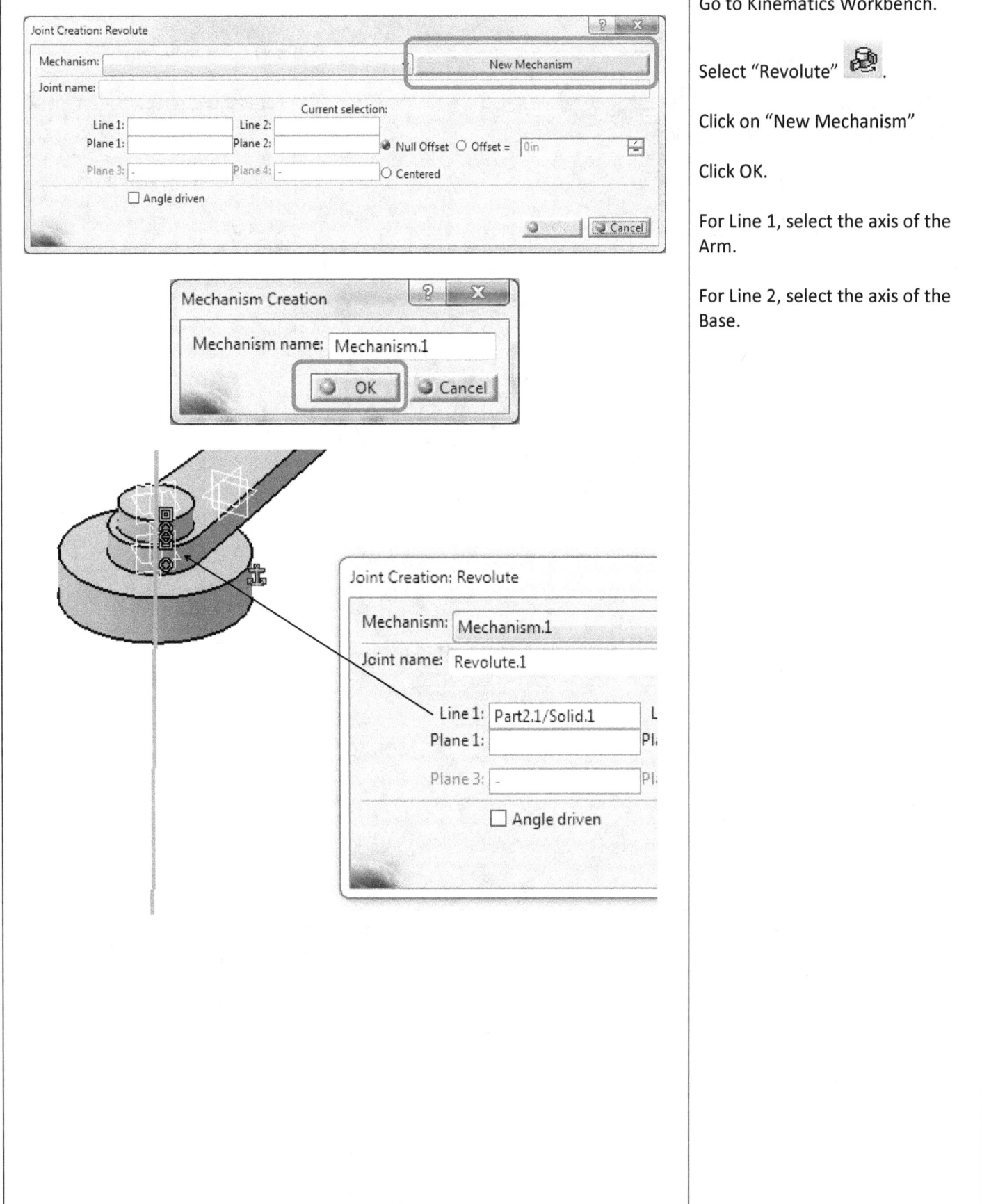

Go to Kinematics Workbench.

Select "Revolute" .

Click on "New Mechanism"

Click OK.

For Line 1, select the axis of the Arm.

For Line 2, select the axis of the Base.

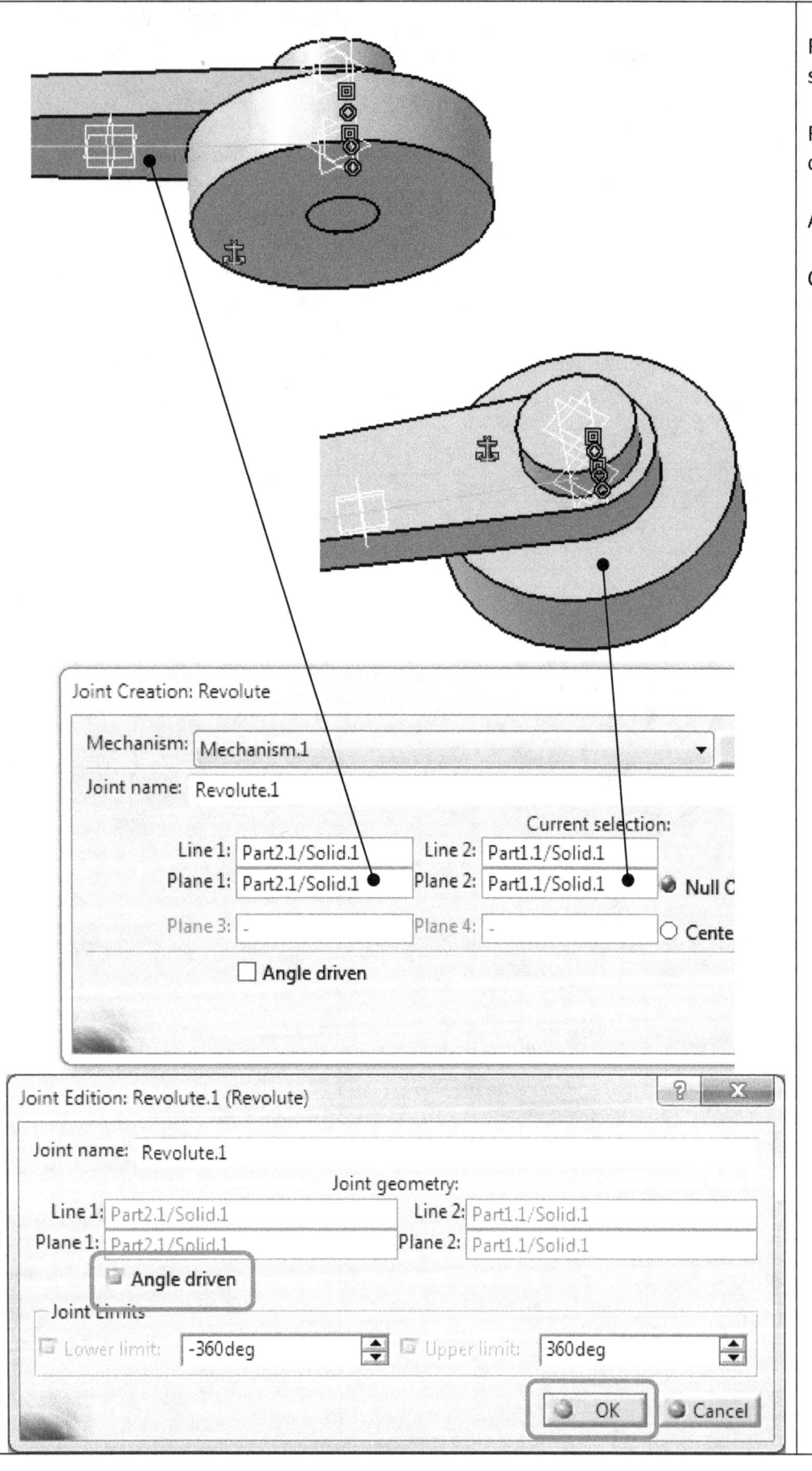

For Plane 1, select the bottom surface of the Arm.

For Plane 2, select the top surface of the Base.

Activate "Angle driven"

Click OK.

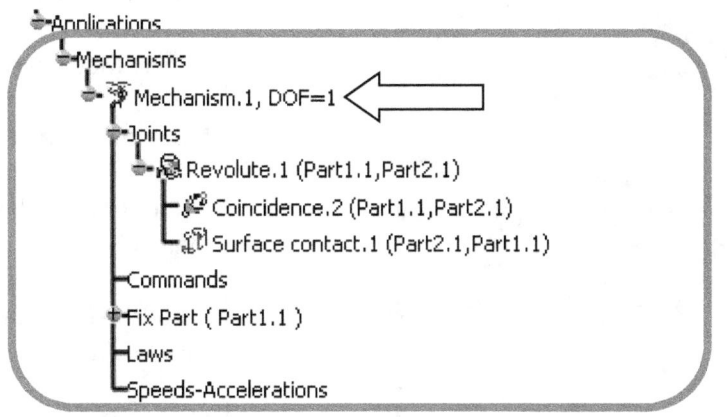

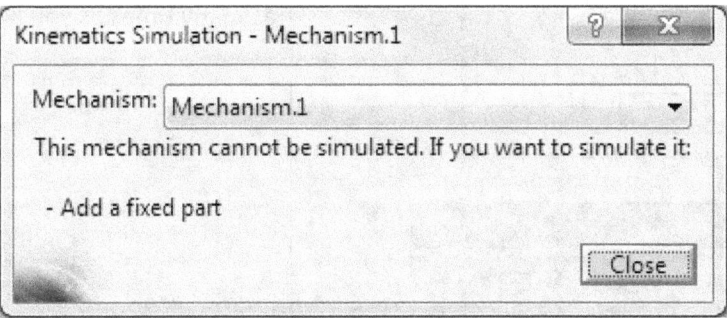

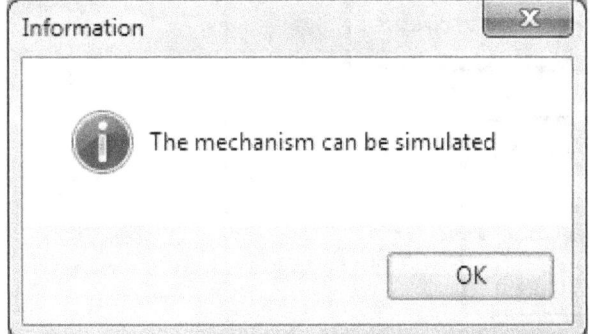

Double click on "Mechanism.1" on the Tree.

An error message might pop up. If so, add the Fix on the Base.

A new message might appear as shown on the left. It is now ready to be simulated.

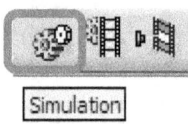

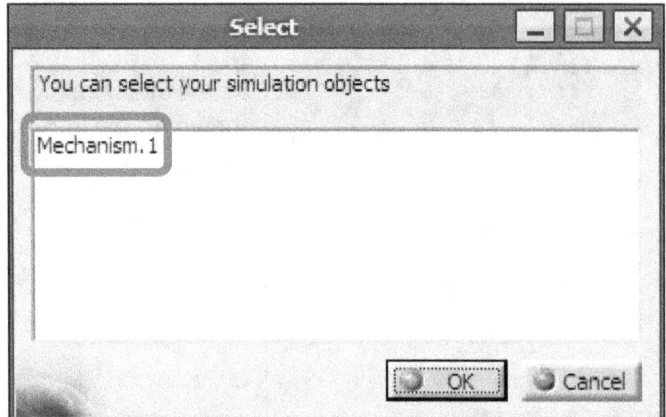

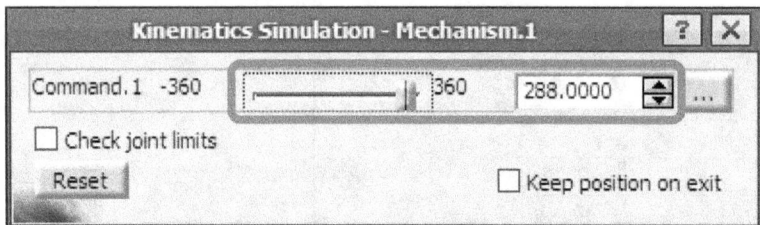

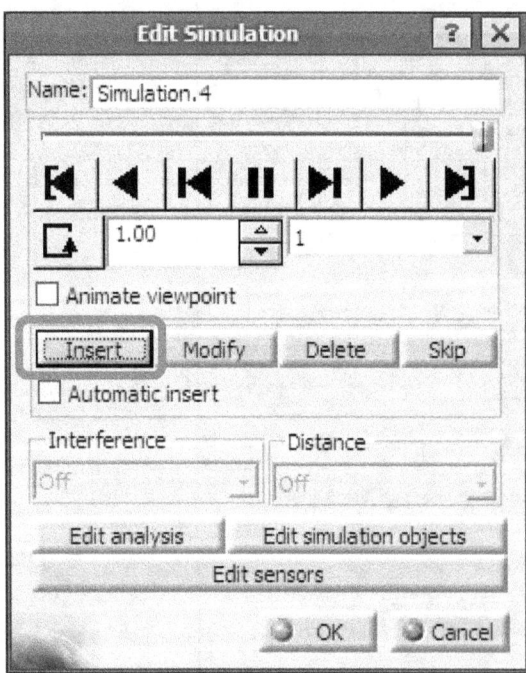

Click the Simulation icon

Select "Mechanism.1" and OK

Move the slider in the Kinematics Simulation box as shown on the left, or enter a specific value manually.

Go to the Edit Simulation box, and click Insert.

Change the value and click Insert as many times as you like.

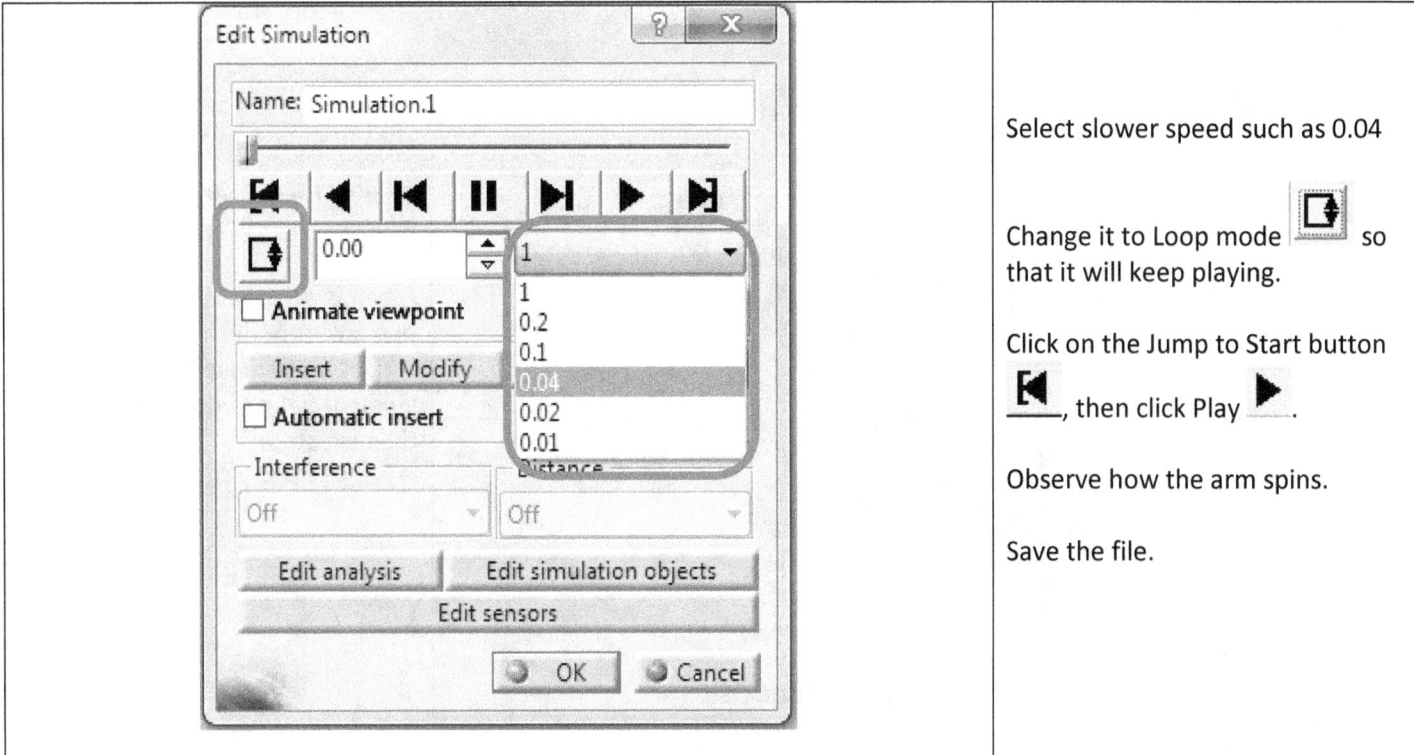

Select slower speed such as 0.04

Change it to Loop mode so that it will keep playing.

Click on the Jump to Start button , then click Play .

Observe how the arm spins.

Save the file.

Assignment

Based on what was covered in this chapter, create your own turbine and add simulation. An example of a turbine is shown below.

Chapter 5: Kinematics II

Screw Joint

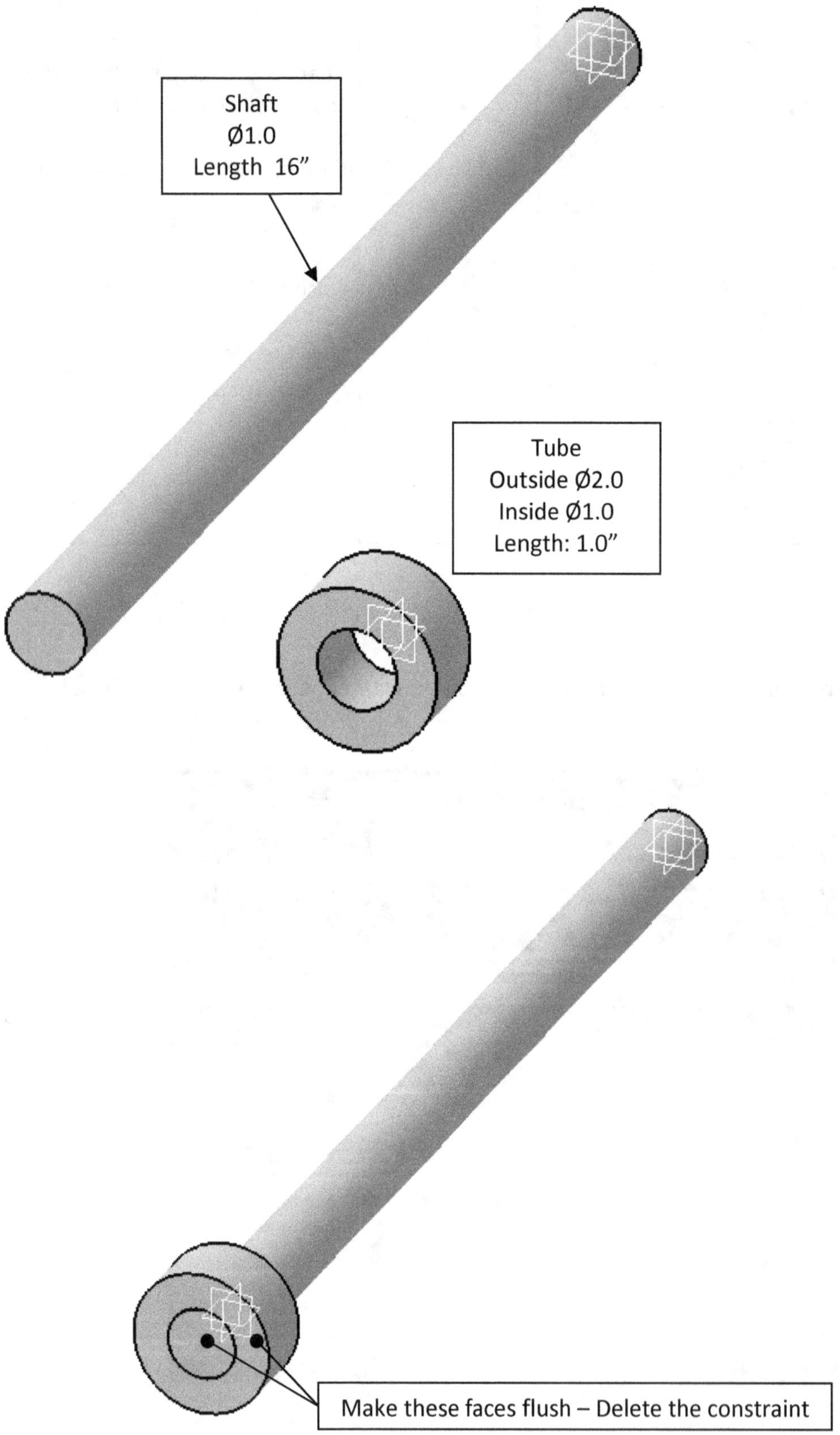

Shaft
Ø1.0
Length 16"

Tube
Outside Ø2.0
Inside Ø1.0
Length: 1.0"

Make these faces flush – Delete the constraint

Create two parts as shown on the left.

Assemble them. Make sure the two faces are flush. (if you use a constraint to make them flush, delete it before you move on to the next step)

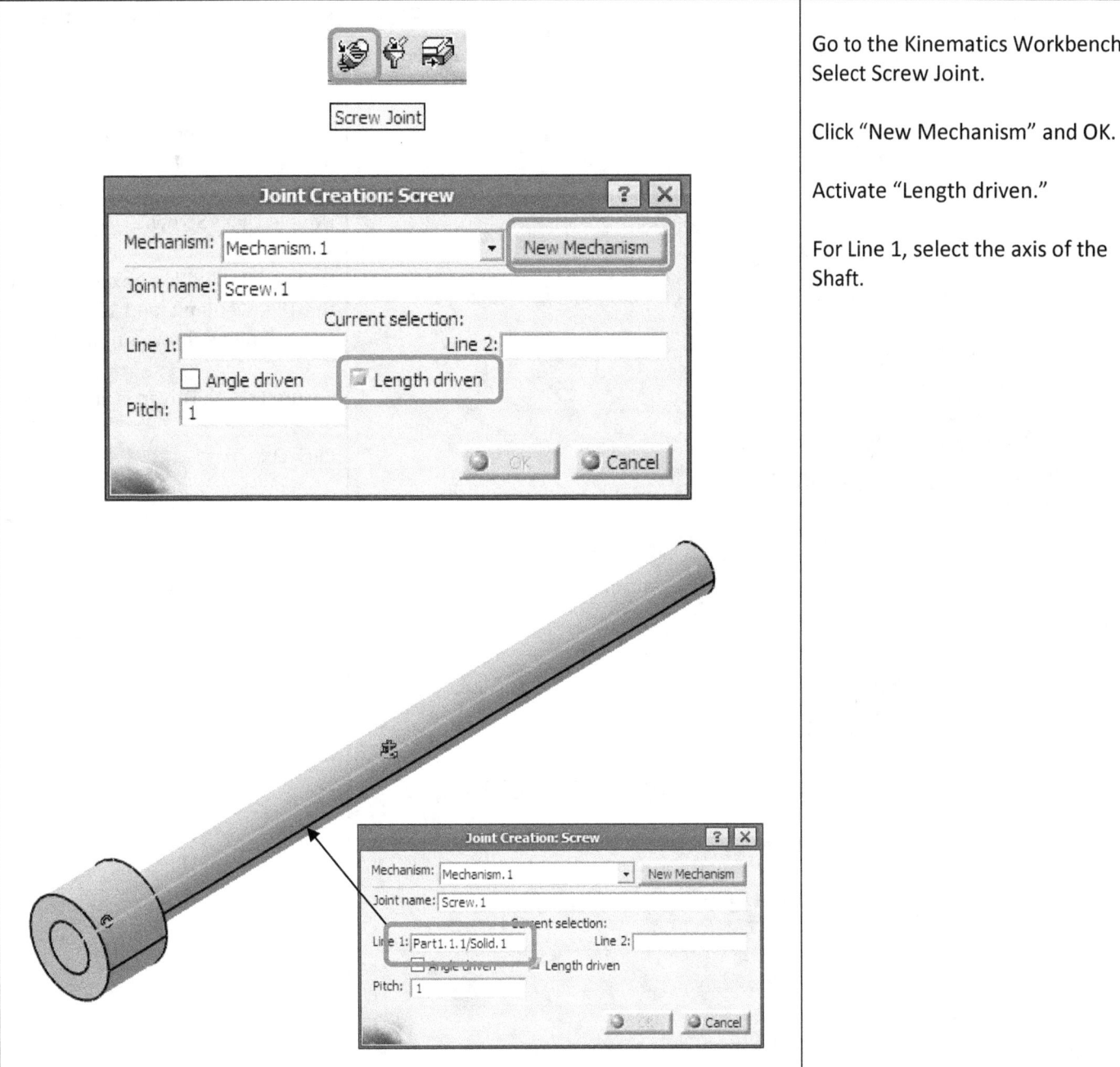

Go to the Kinematics Workbench. Select Screw Joint.

Click "New Mechanism" and OK.

Activate "Length driven."

For Line 1, select the axis of the Shaft.

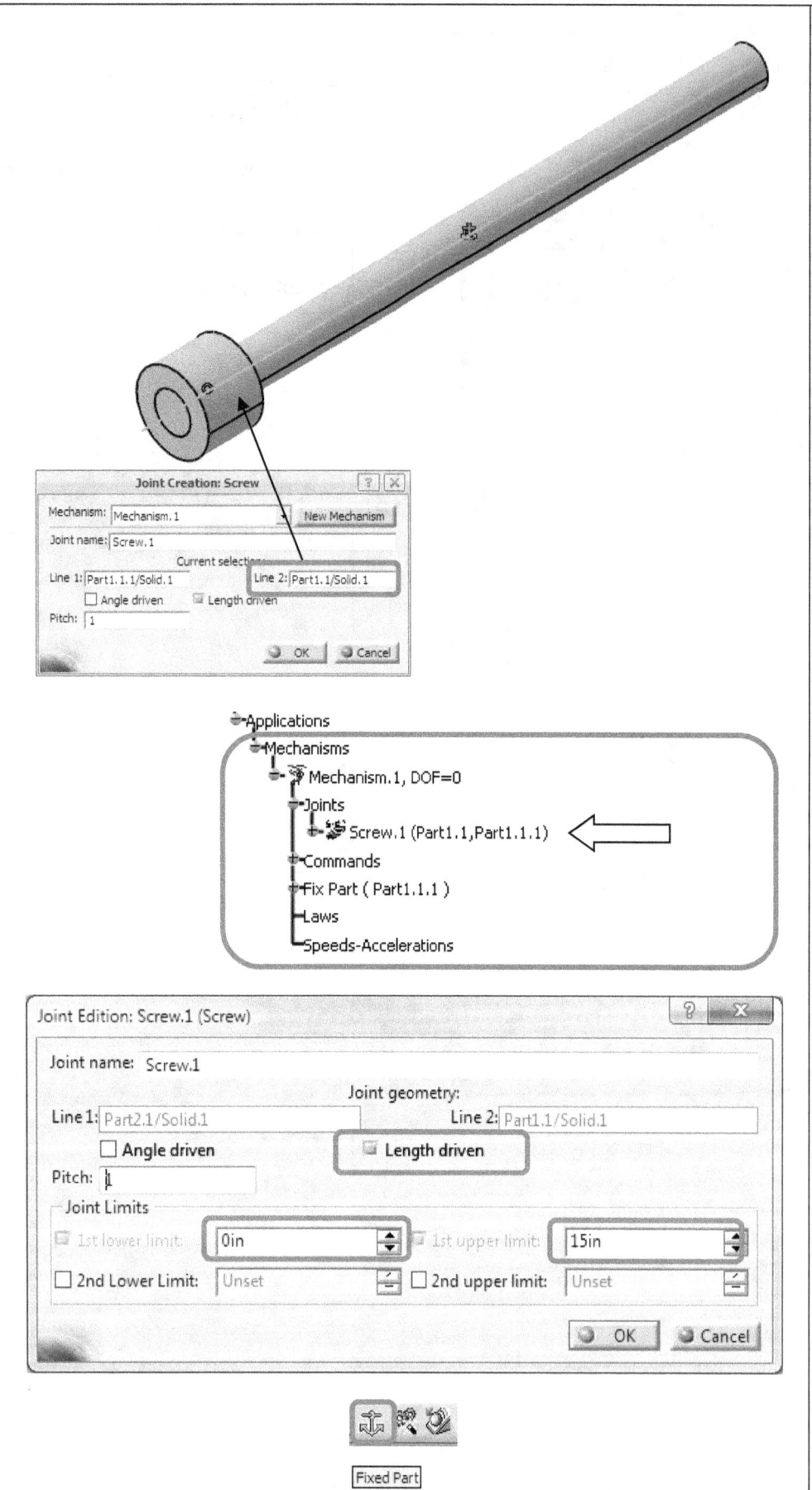

For Line 2, select the axis of the Tube.

Notice "Mechanisms" was created on the tree.

Double click "Screw.1"

On the Joint Edition box, enter 0 in the 1st lower limit, and 15 in the 1st upper limit as shown on the left.

Ensure Length Driven is activated.

Click OK.

Add "Fix" on the Shaft.

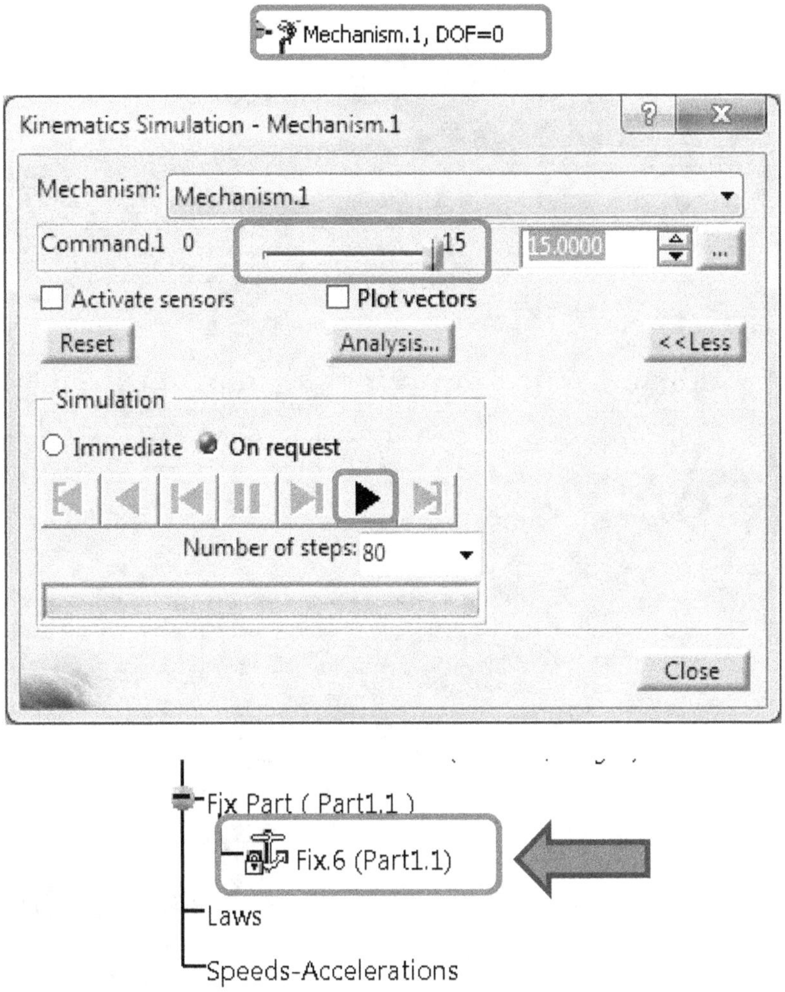

Double click on "Mechnism.1" on the tree.

Move the slider all the way to the right as shown on the left.

Click Play to observe the simulation.

Save the file.

Now, delete the Fix on the Shaft. Add the Fix to the Tube.

Notice how it simulates.

Assignment

Based on what was covered, create a C-clamp and add simulation. An example of a C-clamp is shown below.

Chapter 6: Kinematics III

Point Curve joint

Recreate each part as shown below.

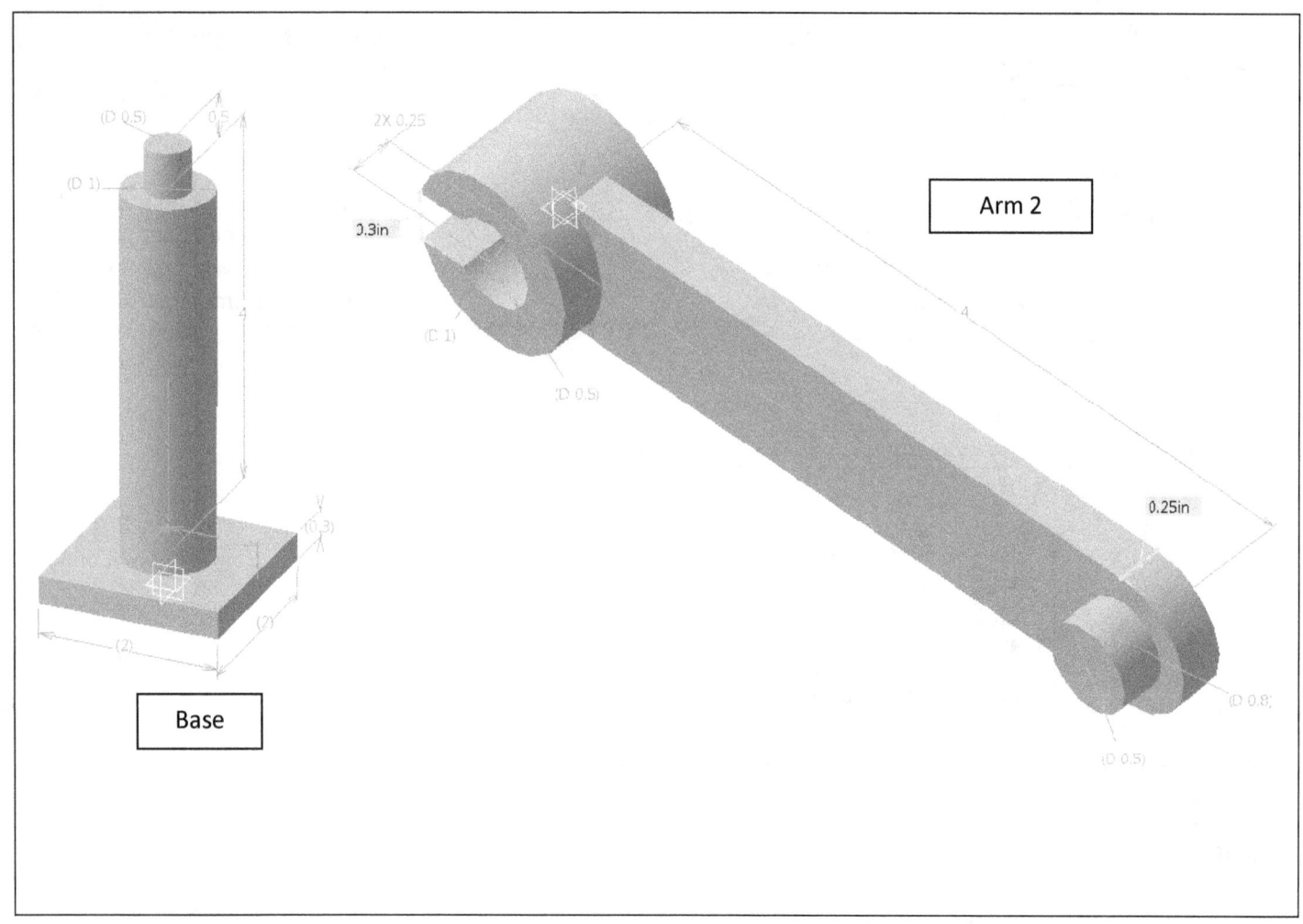

Base

Arm 2

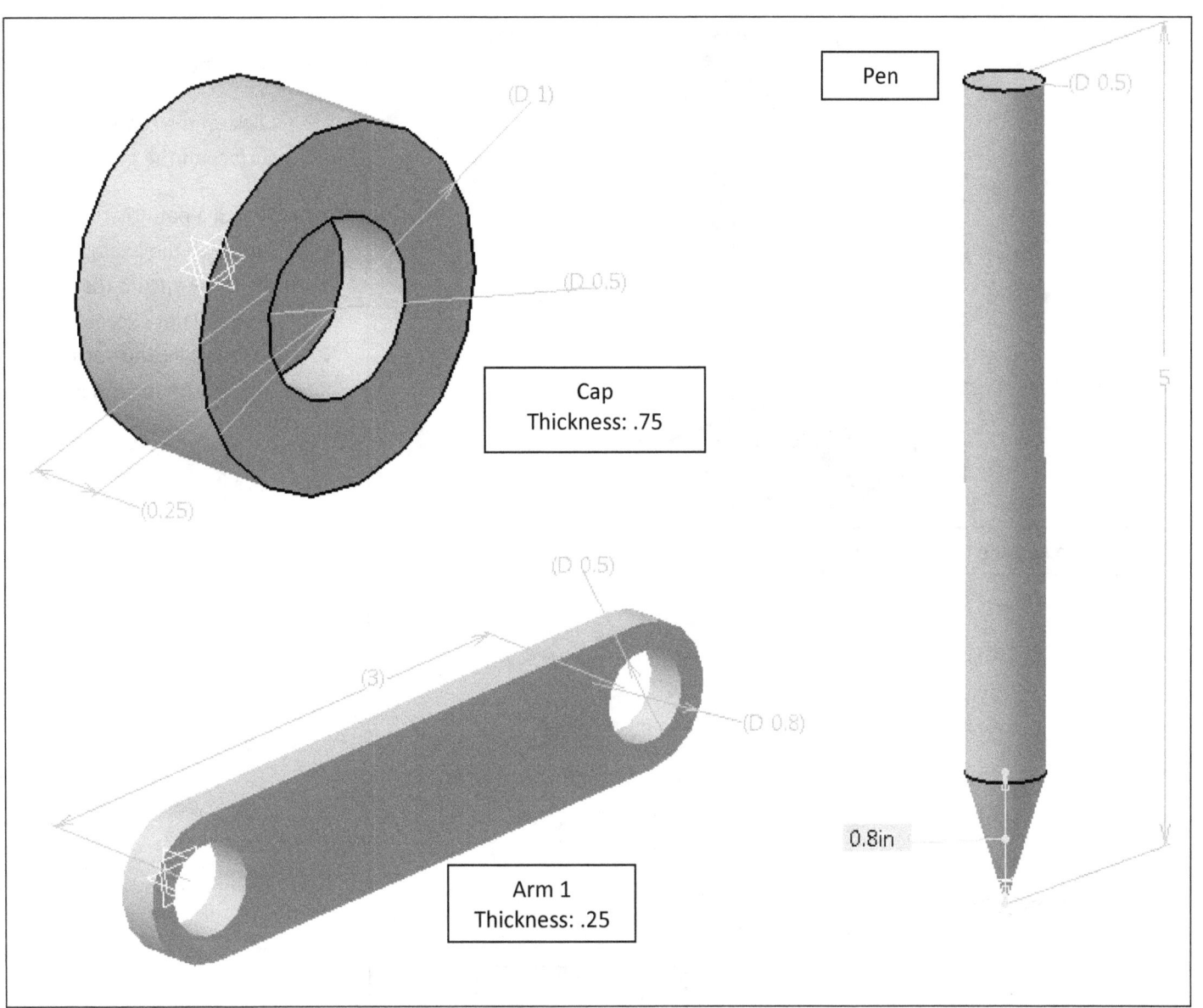

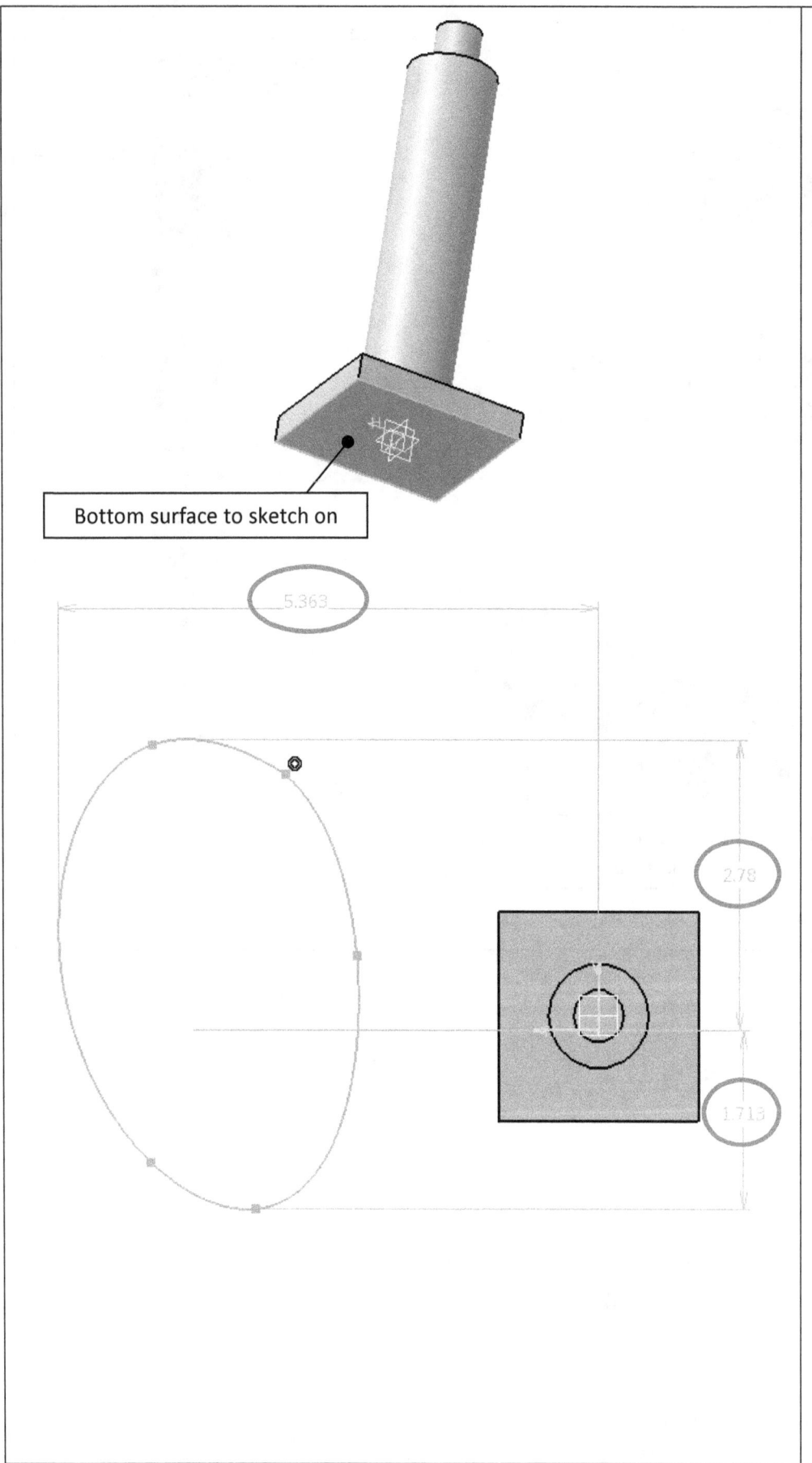

1. Open the "Base."
2. Click on the bottom surface of the plate.
3. Sketch a path that is similar to the example on the left – ensure that the path will not exceed 7" horizontally and vertically--the maximum arm length is 7".
4. Save it.

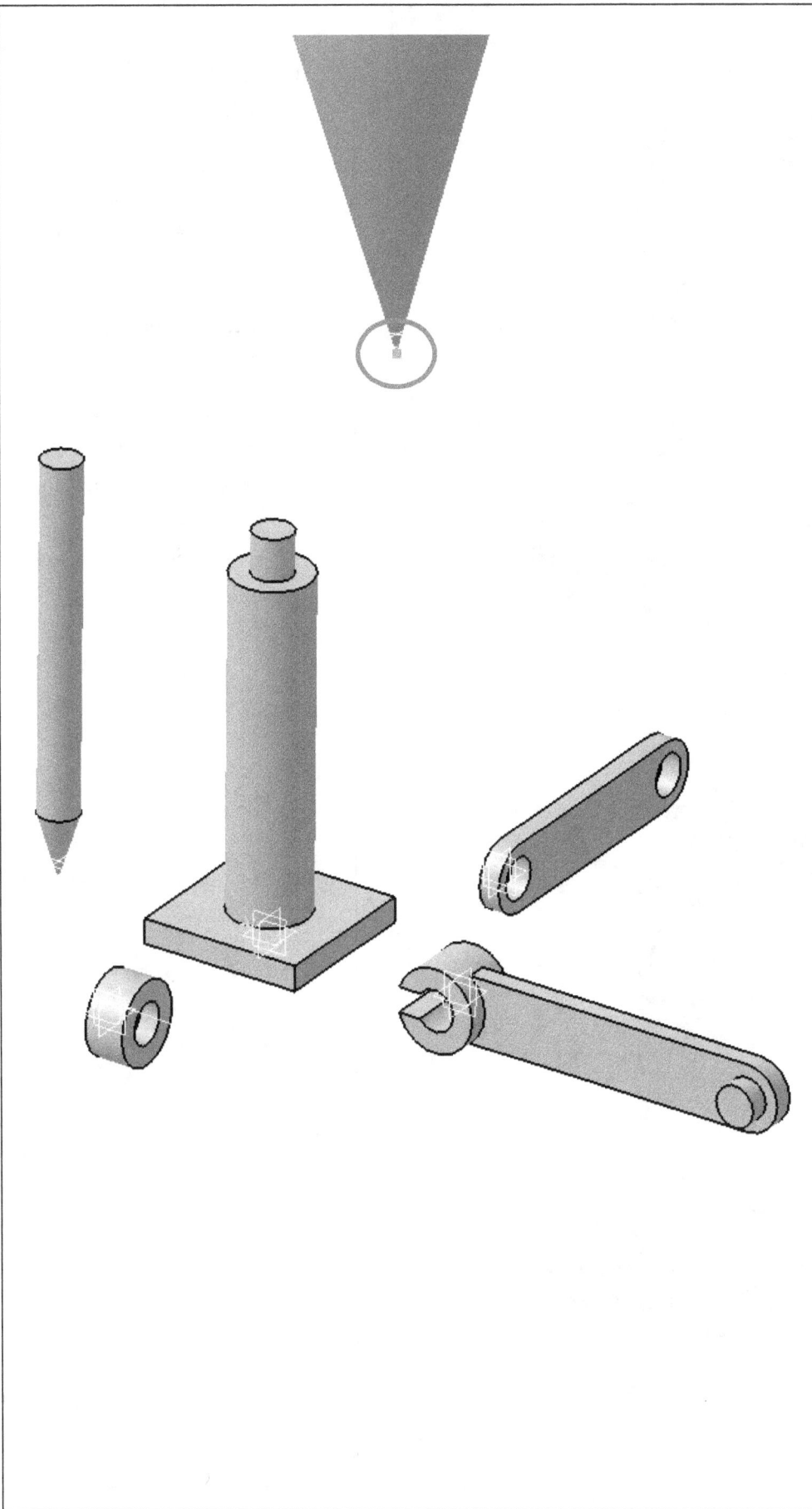

5. Now open the "Pen" and create a point on the tip of it as shown.

6. Save it.

7. Go to Start, Mechanical Design, and select Assembly Design.

8. Insert all parts.

9. Use the Compass and move each part away from each other as shown on the left.

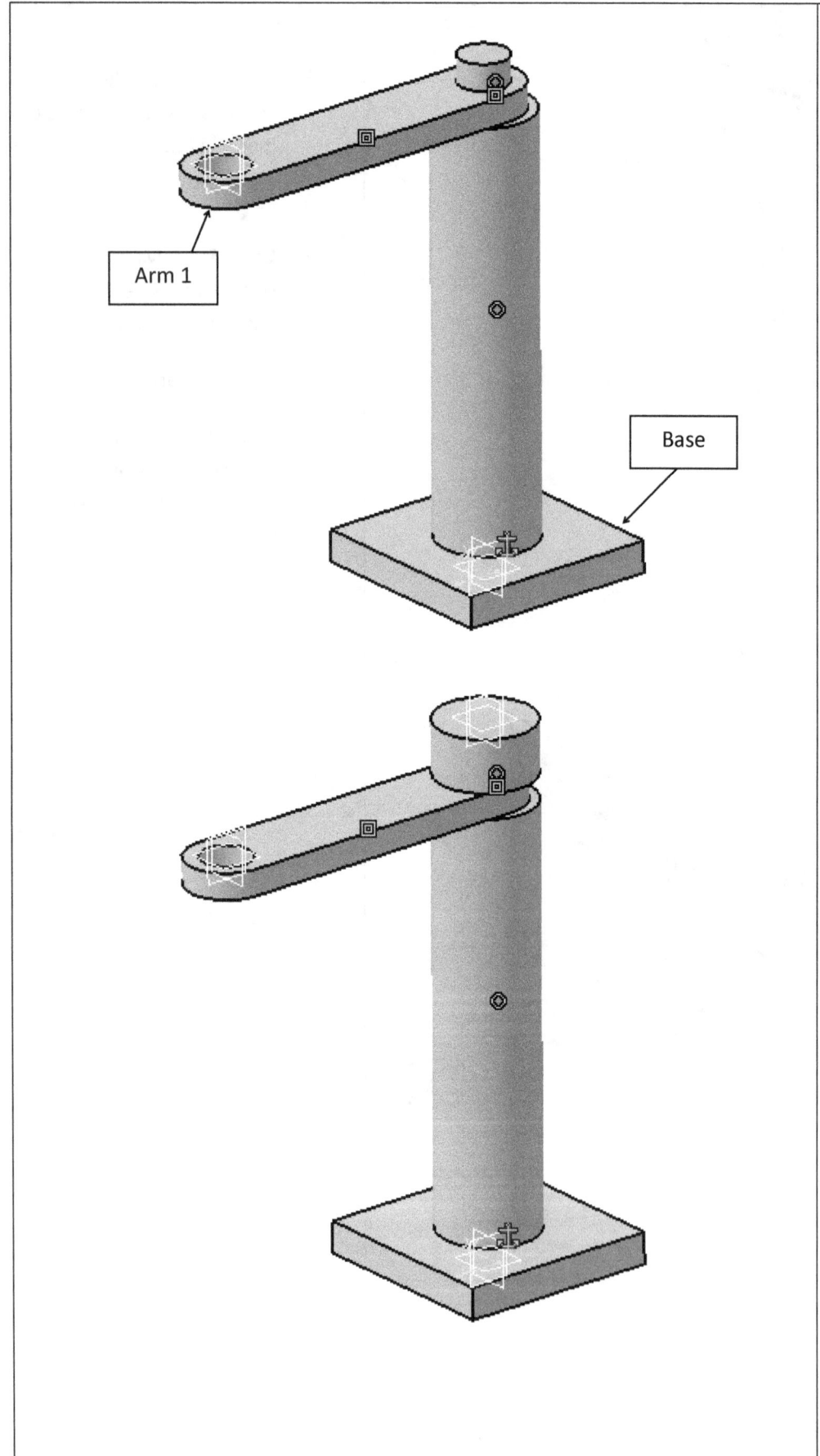

10. Add the Fix ⚓ on the Base.

11. Use the Constraints toolbar and assemble Arm 1 to the Base as shown on the left.

12. Assemble the Cap as shown.

13. Delete the constraints just applied on the Cap – this will ensure that the Cap won't move when it simulates.

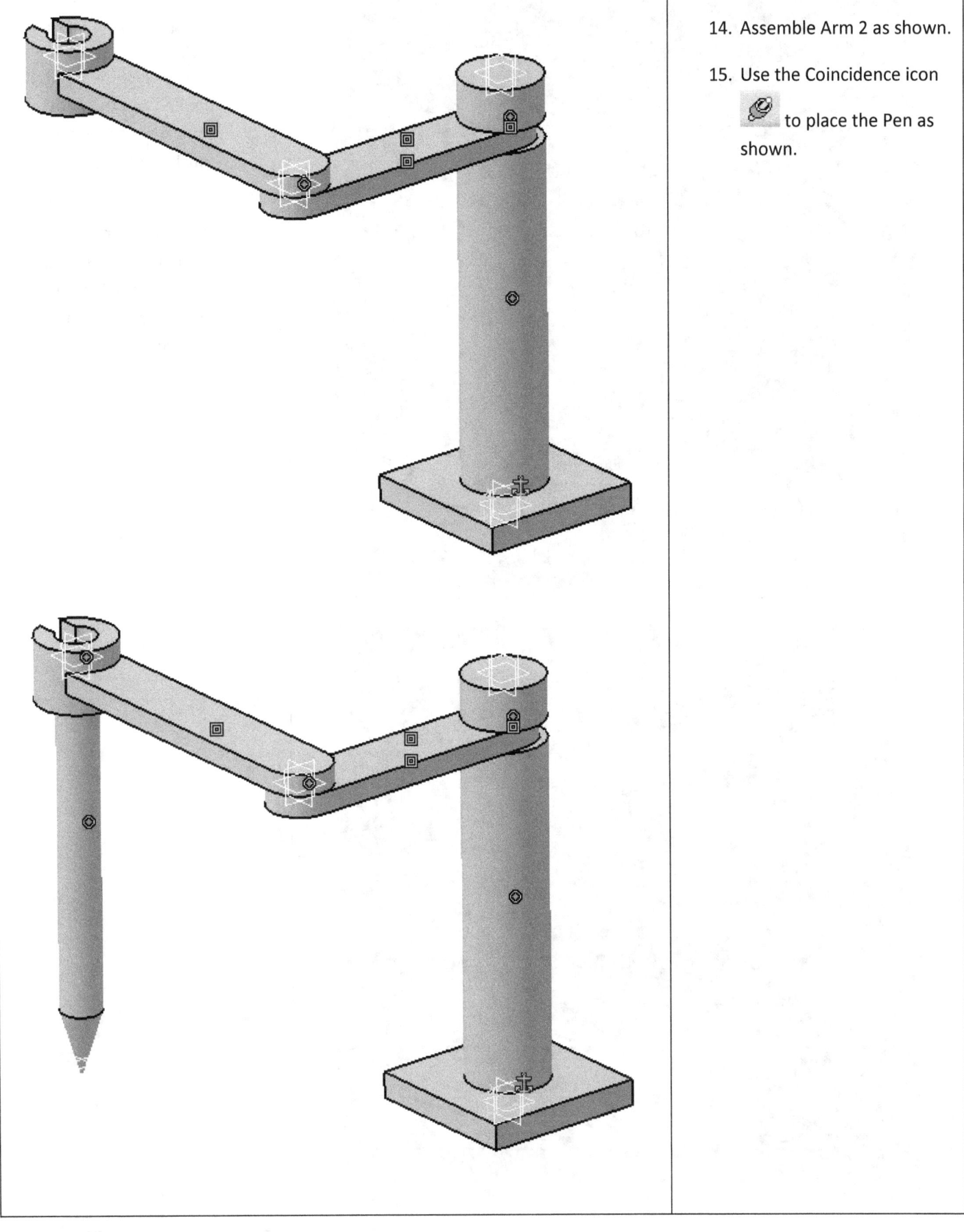

14. Assemble Arm 2 as shown.

15. Use the Coincidence icon to place the Pen as shown.

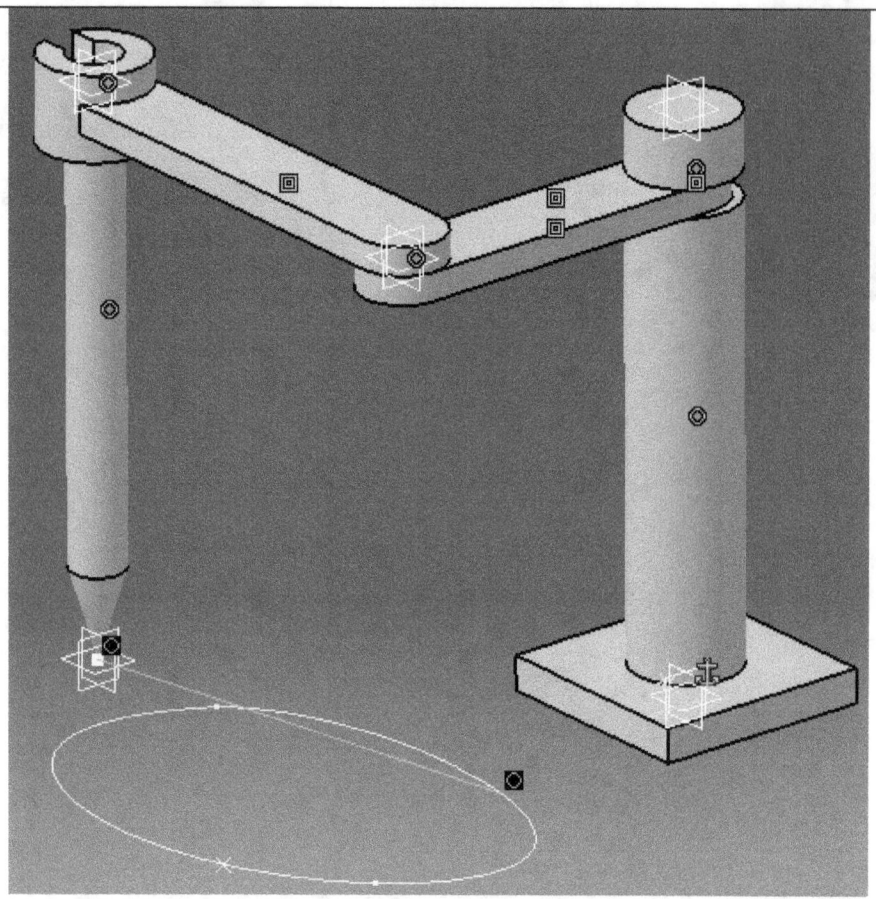

16. Use the Coincidence icon ![icon] select the point of the Pen and the path – this causes the Pen to touch the path.

17. Delete the constraint just applied to the Pen and the path – this will ensure that the Pen moves on the path later.

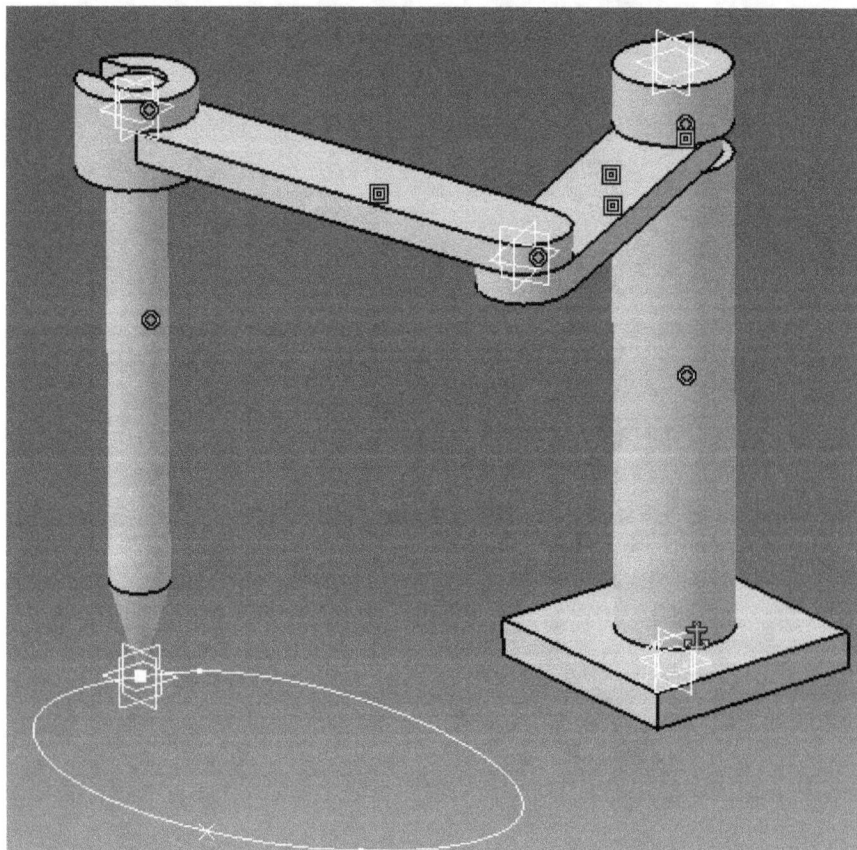

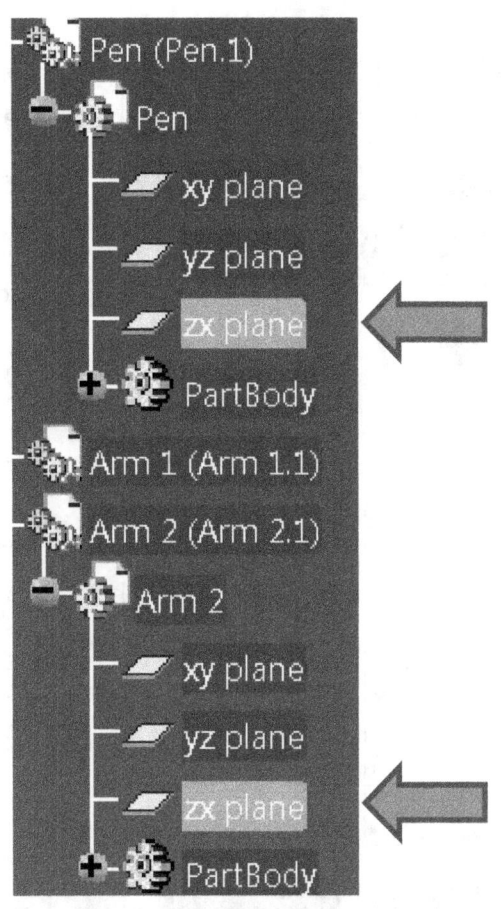

18. Use the Angle Constraint icon and select the ZX planes of the Arm 2 and the Pen – this ensures that the Pen will not spin when it simulates.

19. The value of angle is unimportant. Click OK.

20. Save this assembly as "Robotic Pen."

21. Go to Start, Digital Mockup, and select DMU Kinematics

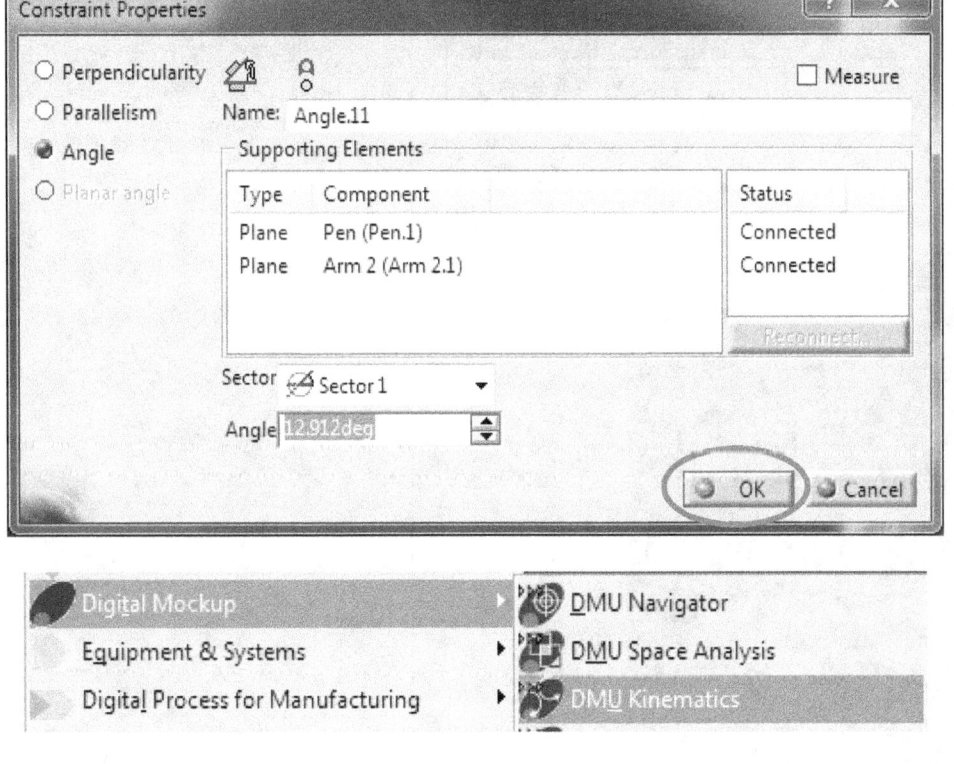

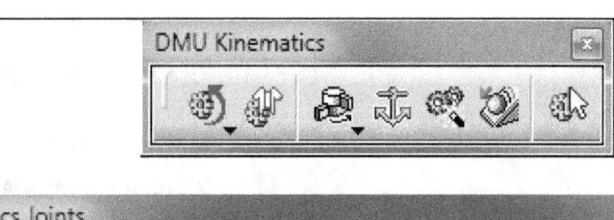

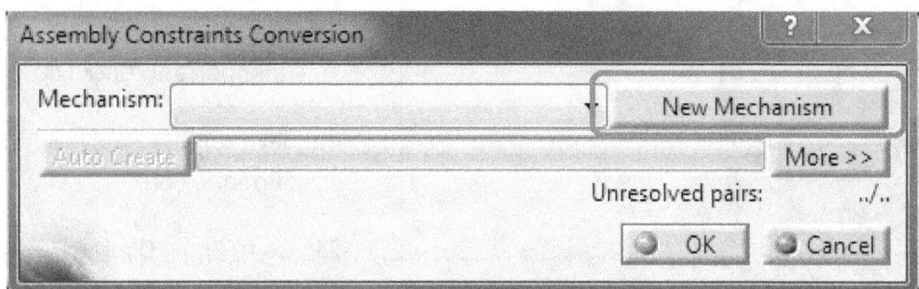
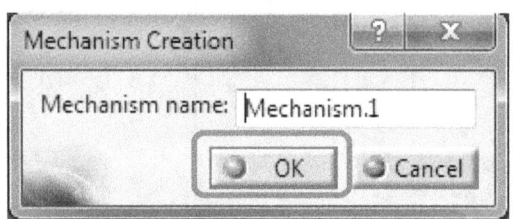
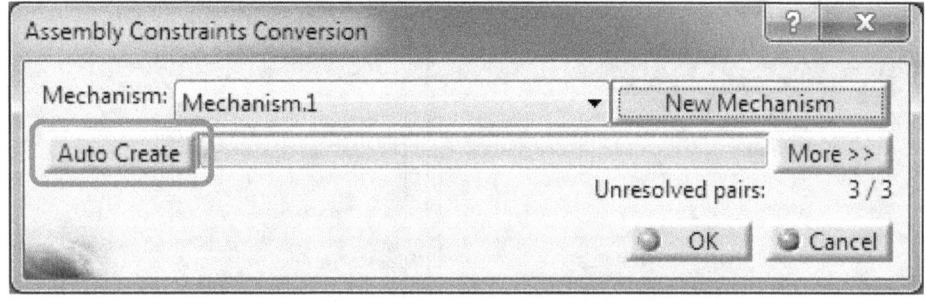
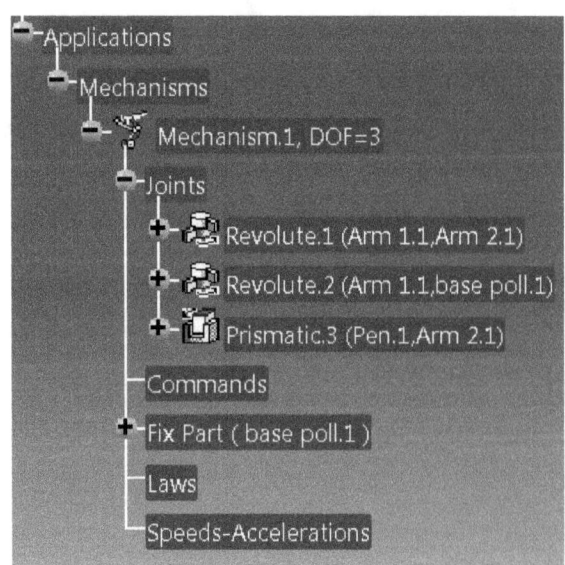

22. Ensure that the two toolbars shown on the left are available.

23. Click on the Assembly Constraints Conversion icon .

24. Click on "New Mechanism."

25. Click OK.

26. Click on the "Auto Create."

27. Applications on the bottom of the Tree should be similar to the example on the left.

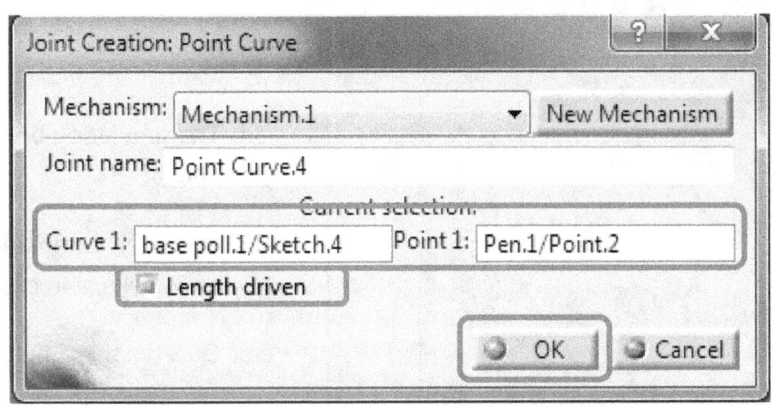

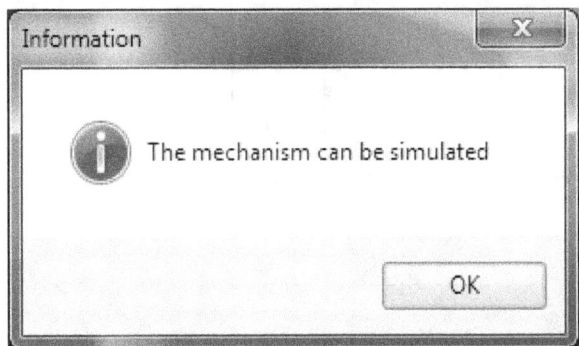

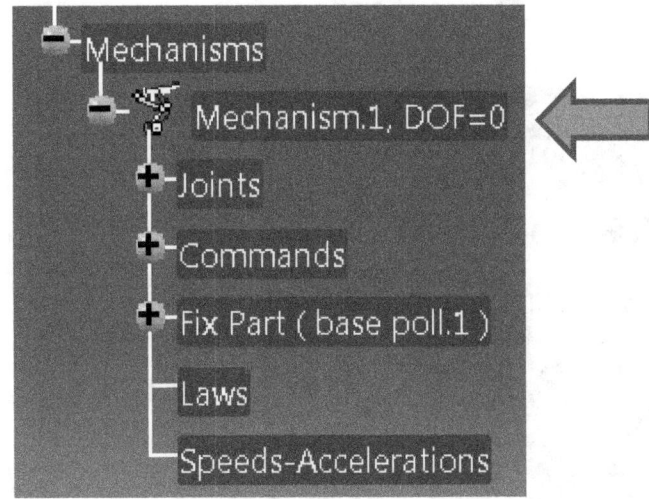

28. Select the Point Curve Joint icon.

29. For Curve 1, select the path.

30. For Point 1, select the point on the Pen.

31. Activate "Length Drive."

32. The 'mechanism can be simulated' message should appear.

33. Double click "Mechanism.1" on the Tree.

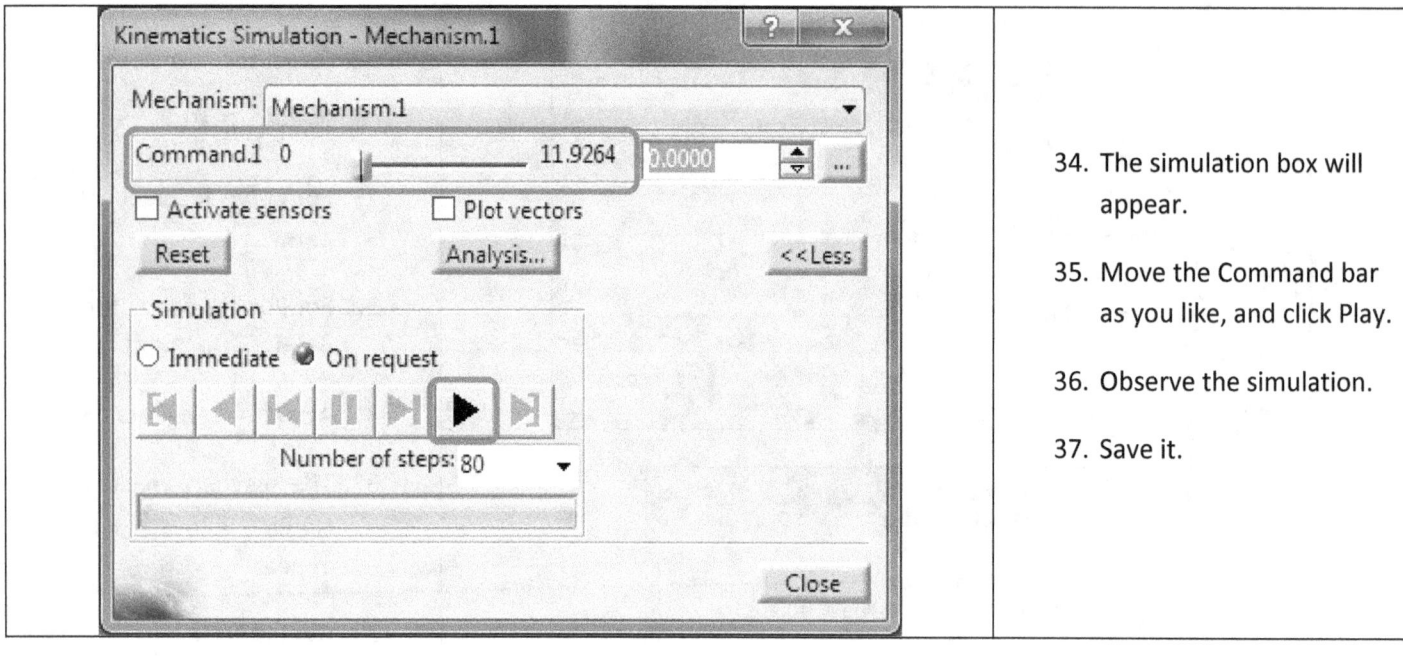

34. The simulation box will appear.
35. Move the Command bar as you like, and click Play.
36. Observe the simulation.
37. Save it.

Assignment

Create a new angles plane on the Base and create a path on it as shown below. Regenerate the simulation. Observe how the Pen moves.

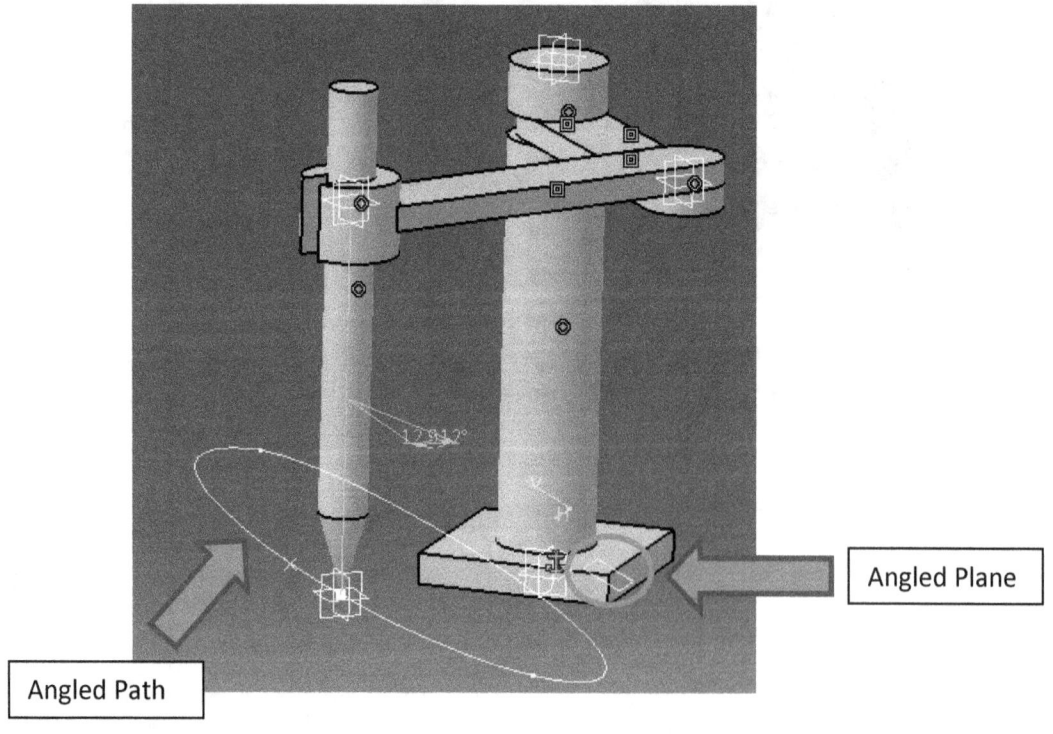

Angled Path

Angled Plane

Chapter 7: Surfacing I

This is an introduction of some basic functions in the Generative Shape Design Workbench.

Toolbar to be used:

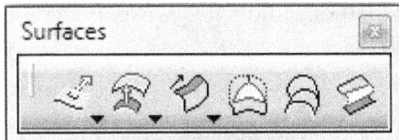

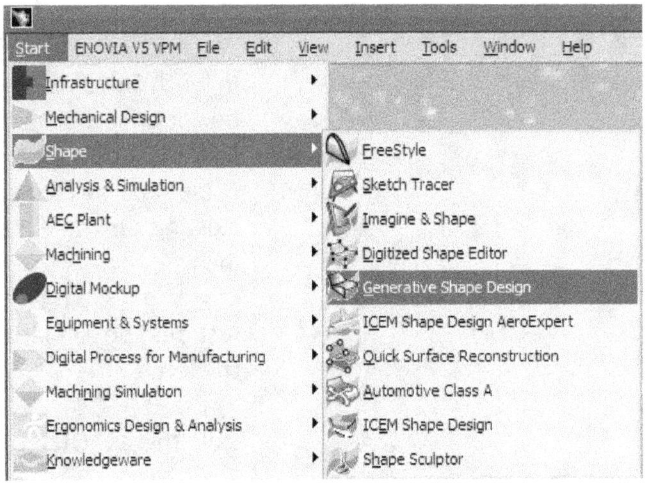

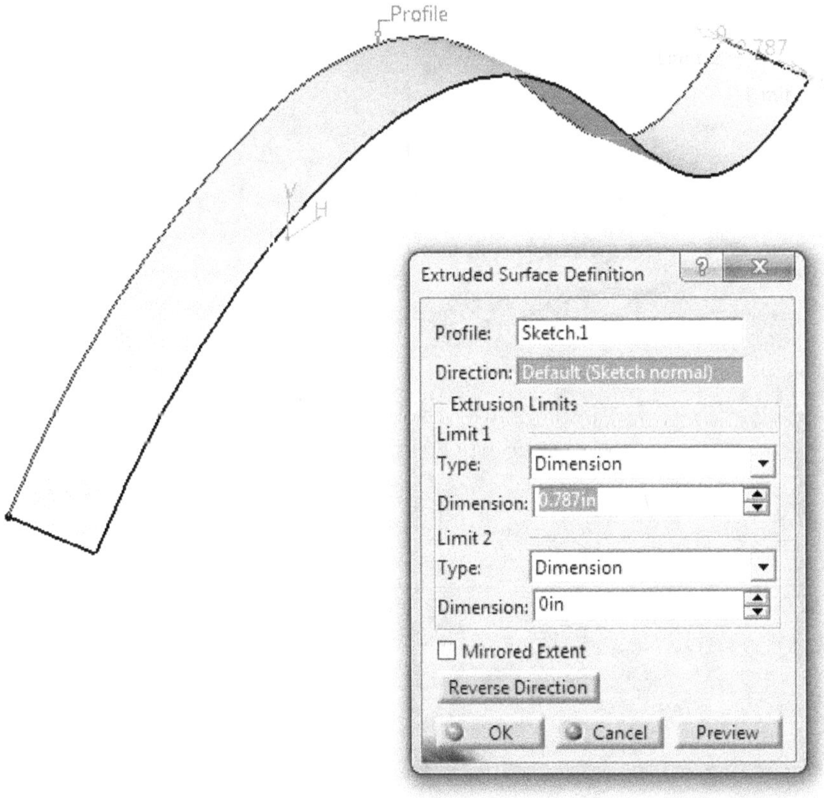

Go to Start, Shape, and then the Generative Shape Design Workbench.

Click OK.

Select the ZX plane, create a simple sketch as similar to the view on the left as possible – dimensions are unimportant.

Exit the Sketcher.

Click the Extrude icon  and select the sketch; observe which direction it extends.

Add values as you like in the Extruded Surface Definition box, or click the Limit 1 or Limit 2 arrows in the graphic area and drag them. The length should change as you drag.

Save your file as "Extrude."

Offset

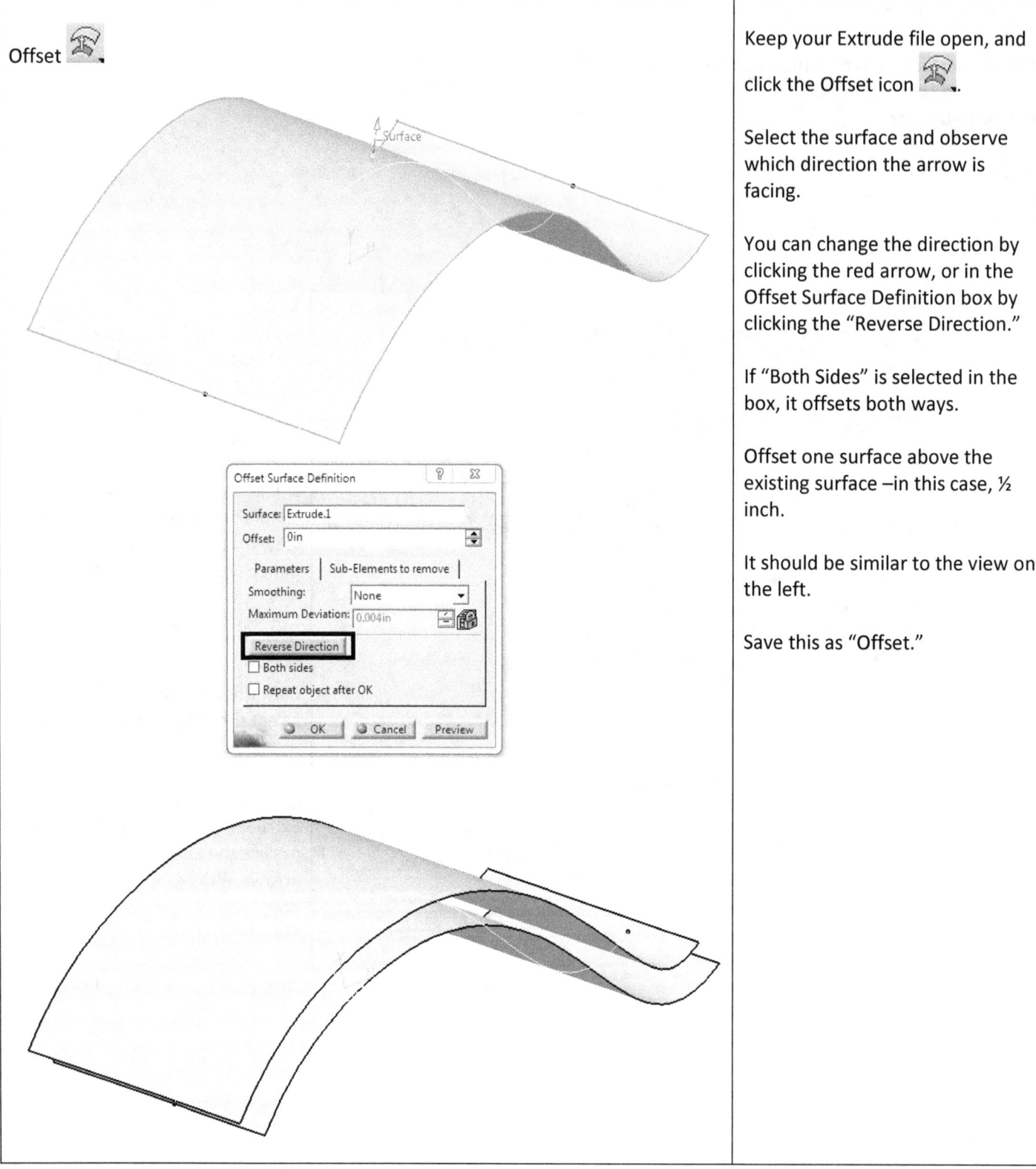

Keep your Extrude file open, and click the Offset icon.

Select the surface and observe which direction the arrow is facing.

You can change the direction by clicking the red arrow, or in the Offset Surface Definition box by clicking the "Reverse Direction."

If "Both Sides" is selected in the box, it offsets both ways.

Offset one surface above the existing surface –in this case, ½ inch.

It should be similar to the view on the left.

Save this as "Offset."

Sweep

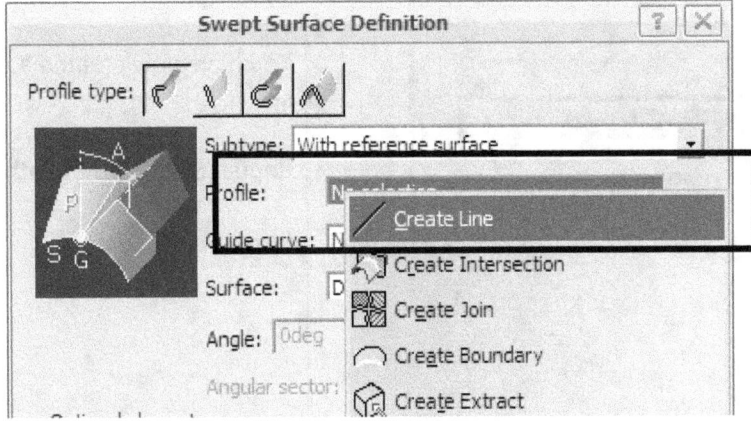

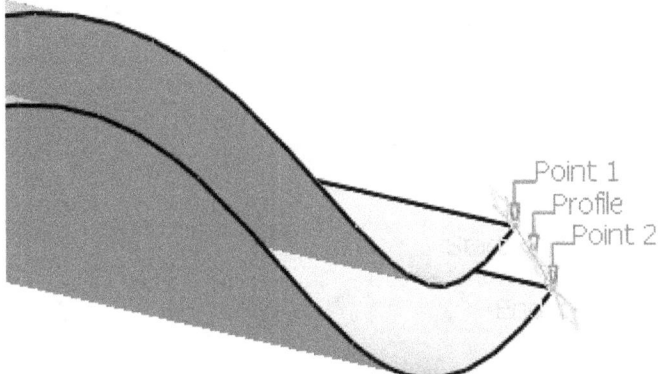

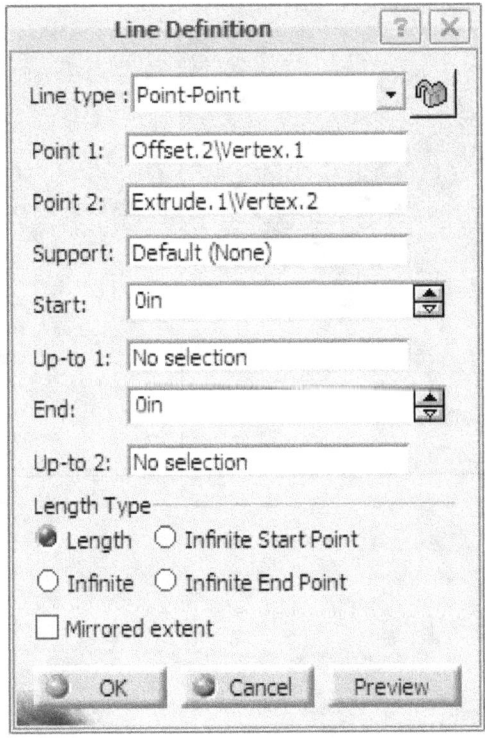

Keep the Offset file open, and click the Sweep icon.

Right click on the Profile area and select "Create line."

Select the corners as shown on the left.

Click OK.

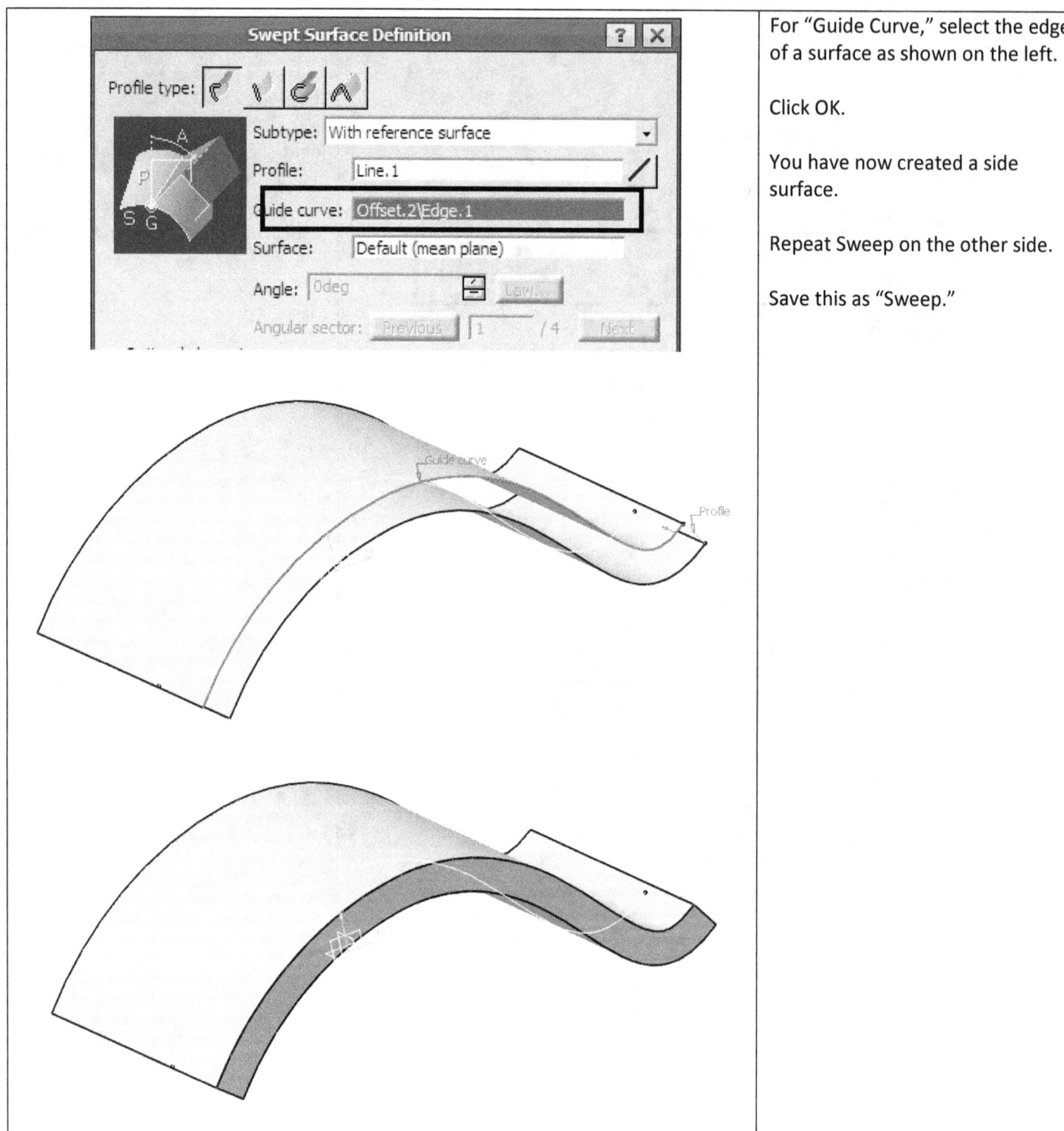

For "Guide Curve," select the edge of a surface as shown on the left.

Click OK.

You have now created a side surface.

Repeat Sweep on the other side.

Save this as "Sweep."

Fill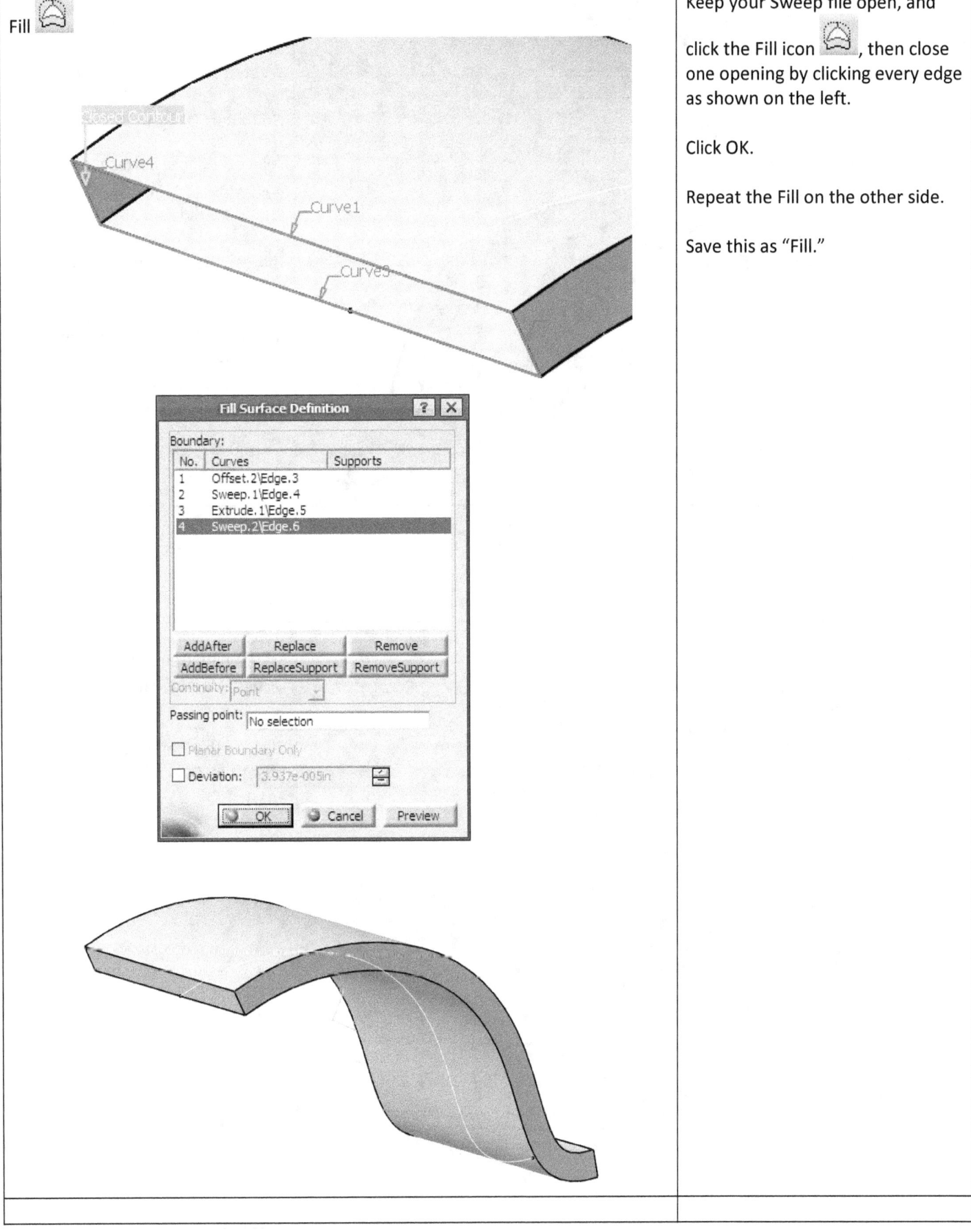

Keep your Sweep file open, and click the Fill icon, then close one opening by clicking every edge as shown on the left.

Click OK.

Repeat the Fill on the other side.

Save this as "Fill."

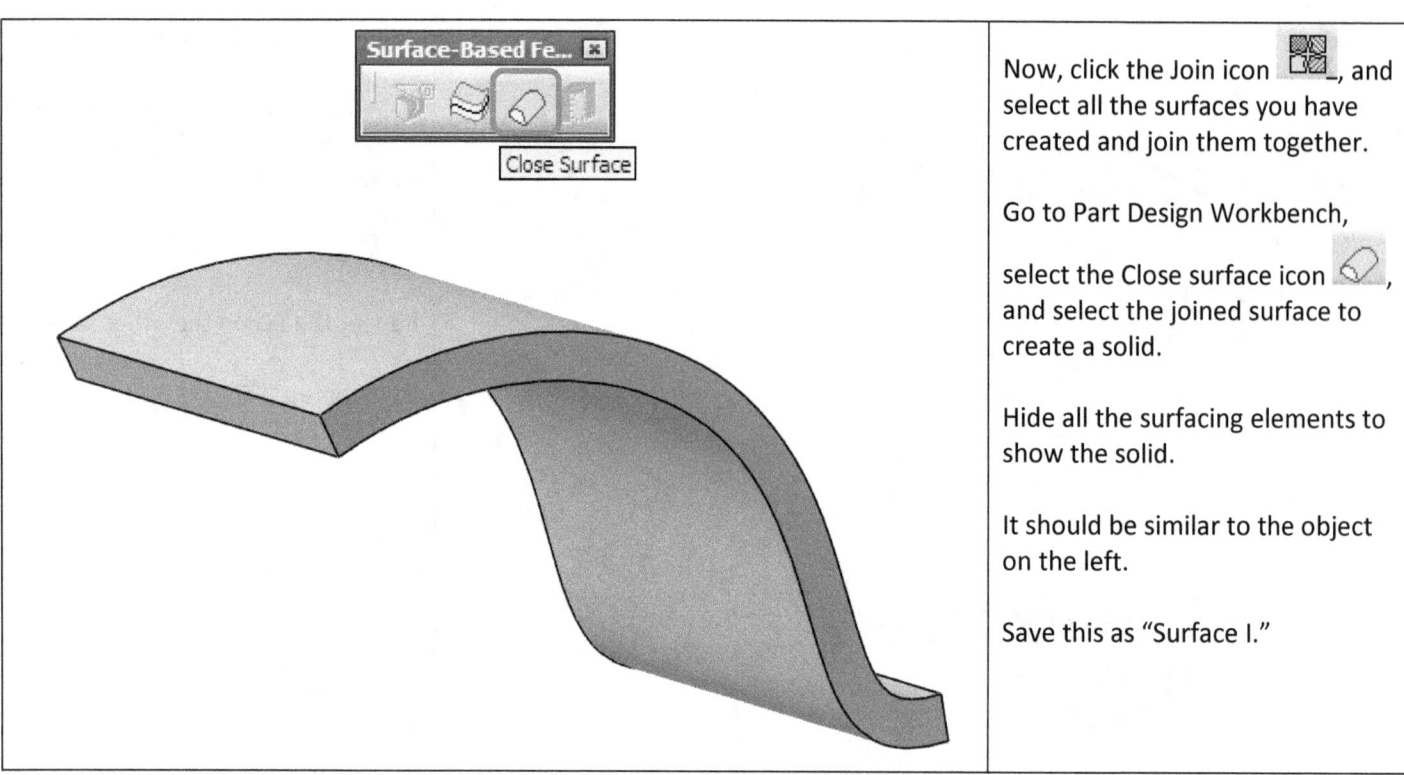

Now, click the Join icon, and select all the surfaces you have created and join them together.

Go to Part Design Workbench, select the Close surface icon, and select the joined surface to create a solid.

Hide all the surfacing elements to show the solid.

It should be similar to the object on the left.

Save this as "Surface I."

Multi-Sections Surface

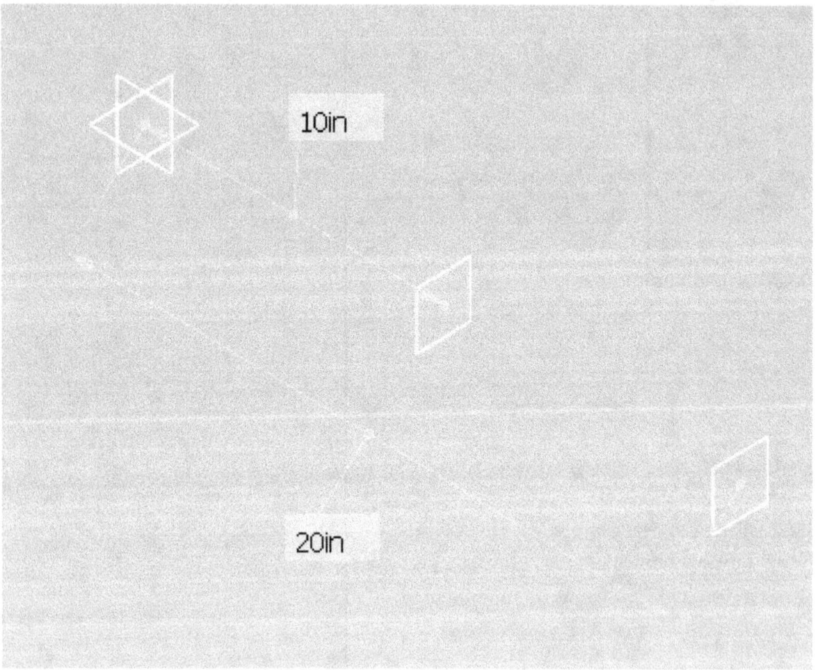

Start a new Generative Shape Design Workbench.

Click on the Plane icon, and offset the ZX plane two times, 10" apart.

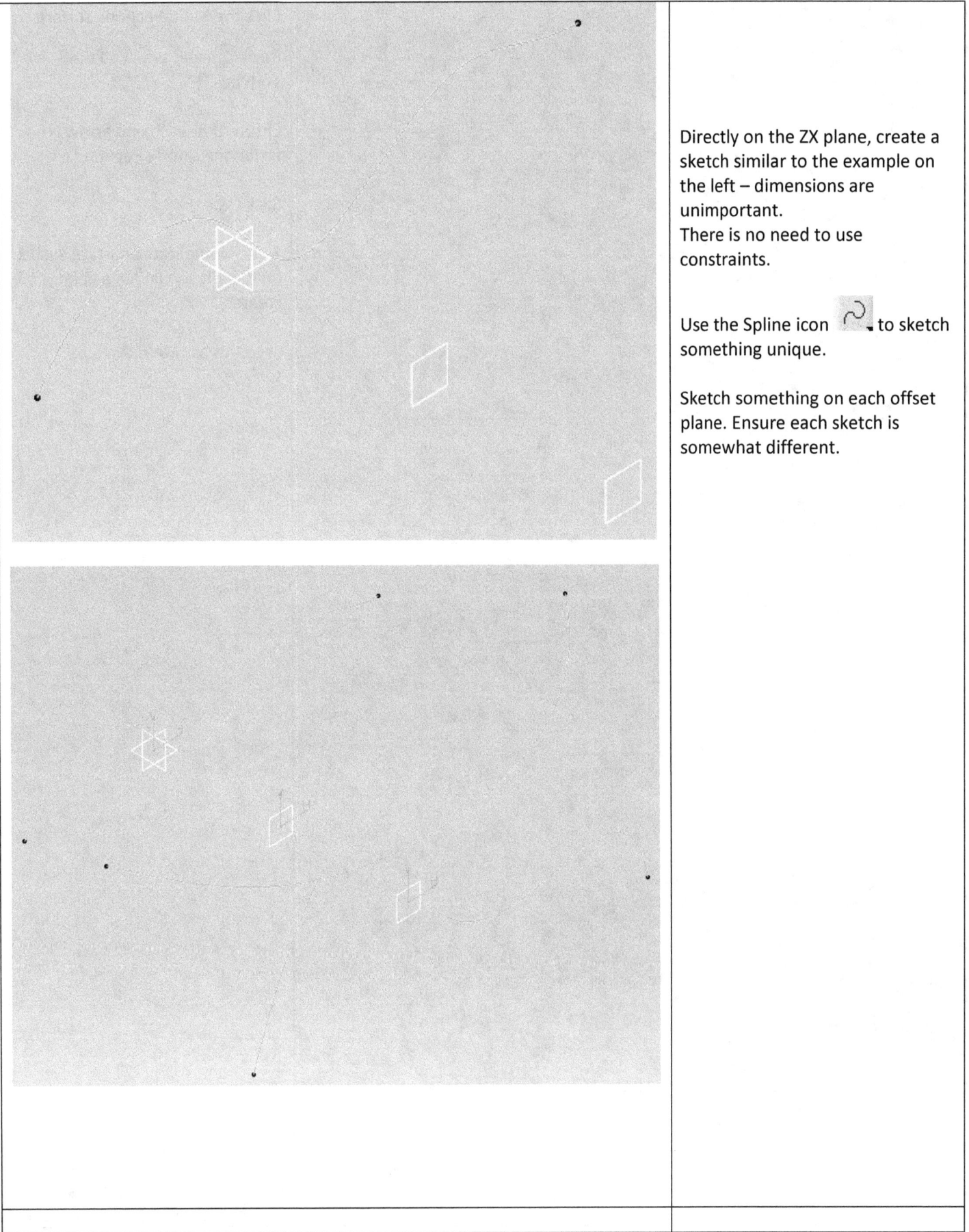

Directly on the ZX plane, create a sketch similar to the example on the left – dimensions are unimportant.
There is no need to use constraints.

Use the Spline icon to sketch something unique.

Sketch something on each offset plane. Ensure each sketch is somewhat different.

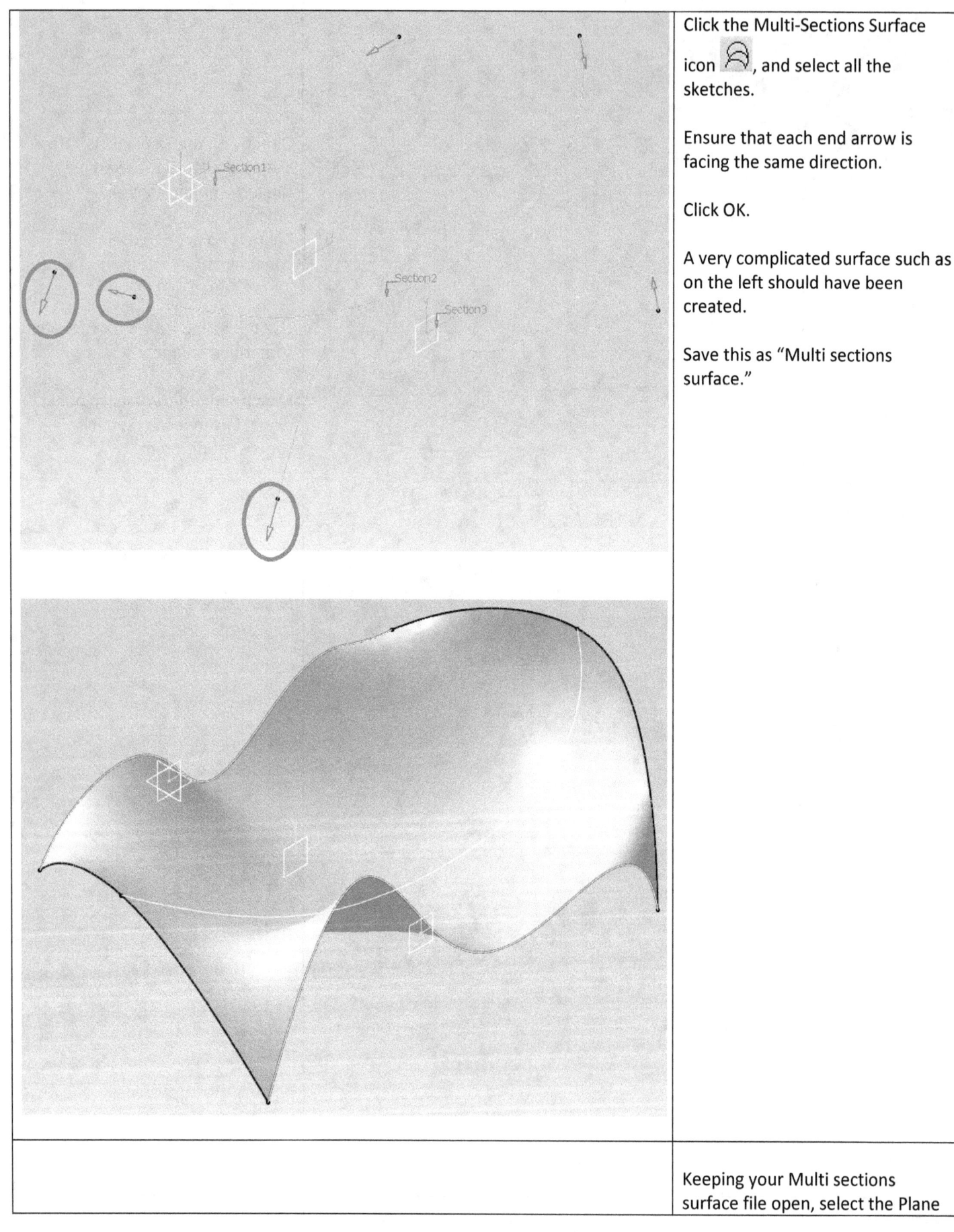

Click the Multi-Sections Surface icon, and select all the sketches.

Ensure that each end arrow is facing the same direction.

Click OK.

A very complicated surface such as on the left should have been created.

Save this as "Multi sections surface."

Keeping your Multi sections surface file open, select the Plane

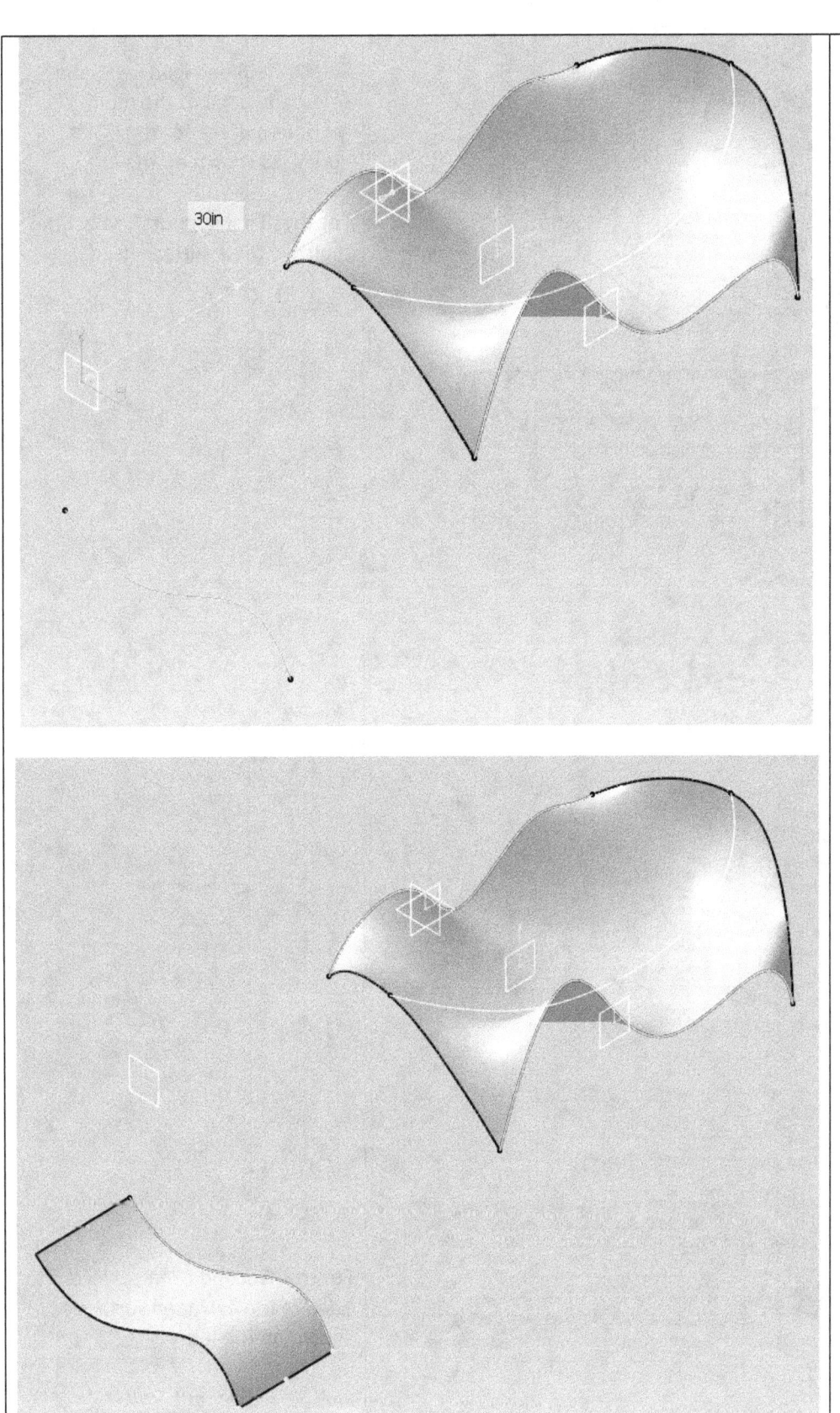

icon , and offset the YZ plane 30" and sketch something on the plane as shown on the left.

Select the Extrude icon , select the sketch you just created, and extrude it 10" as shown on the left.

Hide all sketches for clarity.

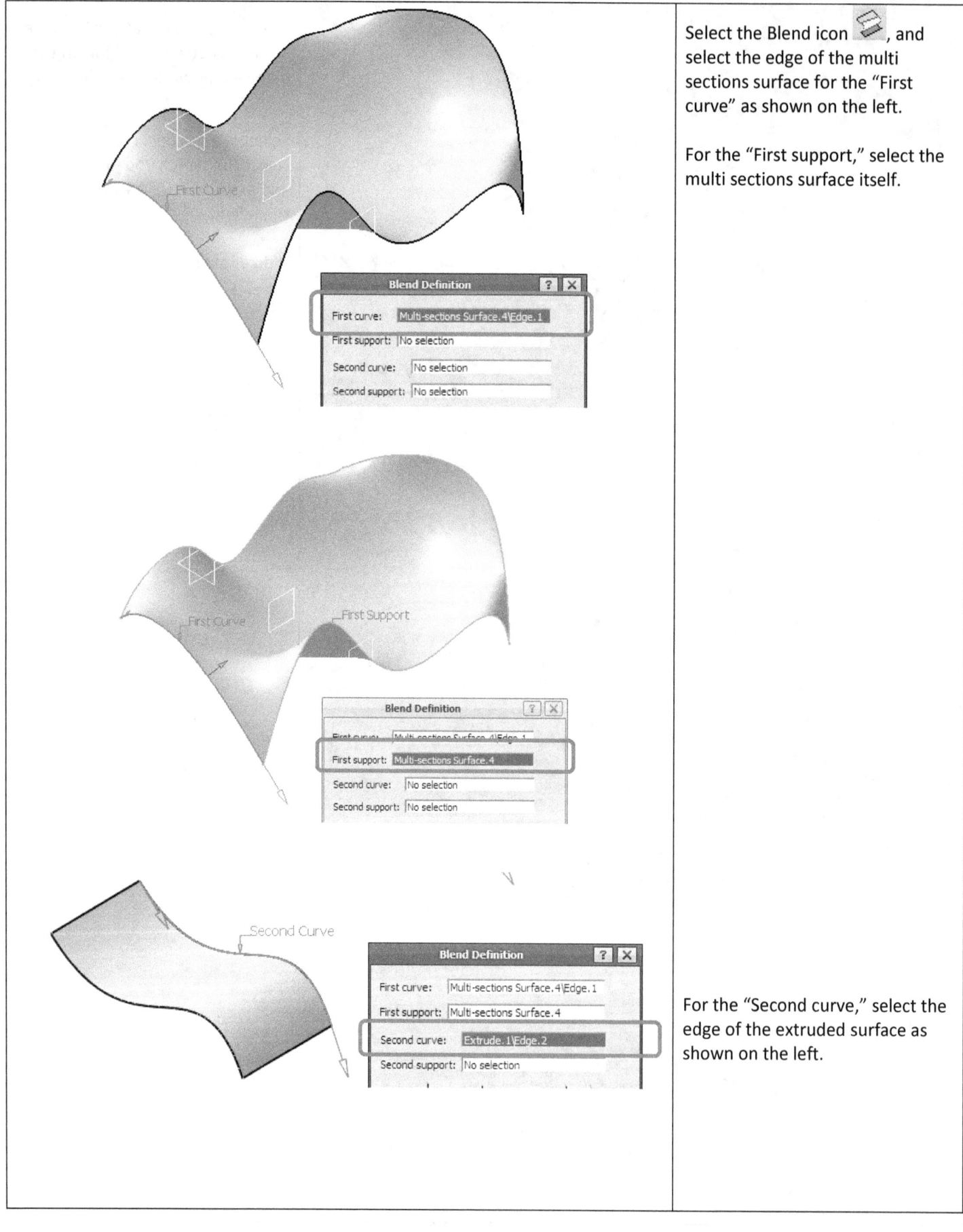

Select the Blend icon, and select the edge of the multi sections surface for the "First curve" as shown on the left.

For the "First support," select the multi sections surface itself.

For the "Second curve," select the edge of the extruded surface as shown on the left.

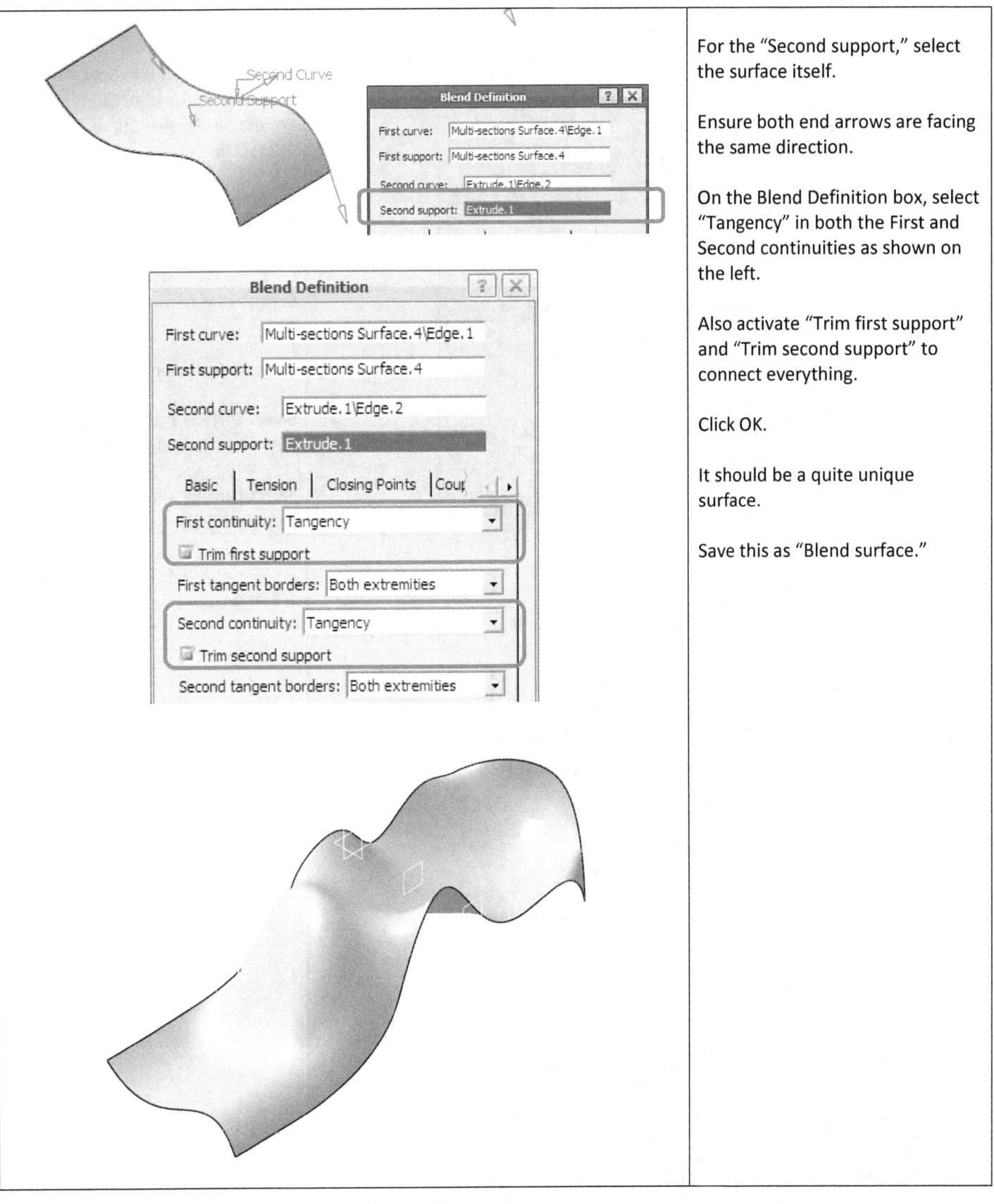

For the "Second support," select the surface itself.

Ensure both end arrows are facing the same direction.

On the Blend Definition box, select "Tangency" in both the First and Second continuities as shown on the left.

Also activate "Trim first support" and "Trim second support" to connect everything.

Click OK.

It should be a quite unique surface.

Save this as "Blend surface."

Chapter 8: Surfacing II

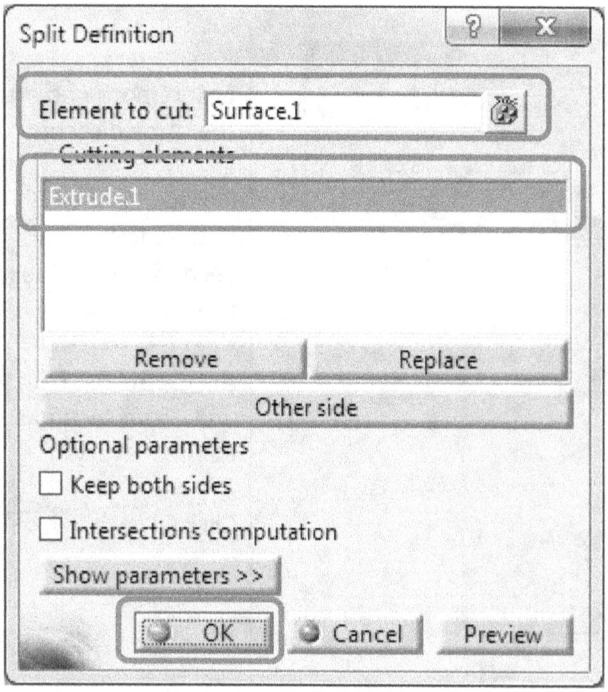

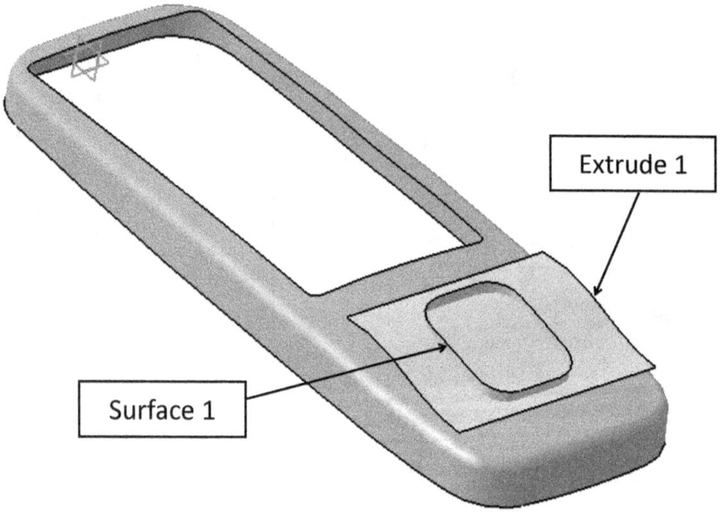

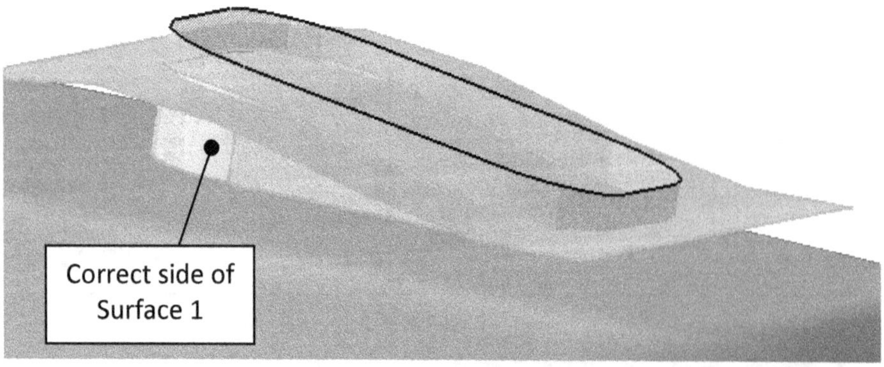

This exercise includes Split, Trim, Disassemble, Untrim, Fillets, Projection, Parallel Curve, and Intersection.

Open the Trimming CATPart. from Class Drive under DGET 4470, within the Surfacing folder.

Set the Unit to mm.

Click the Spilt icon .

Select Surface 1 for "Element to cut" in the Split Definition box.

Select Extrude 1 for "Cutting elements."

Ensure the correct side of Surface 1 remains.

Click OK.

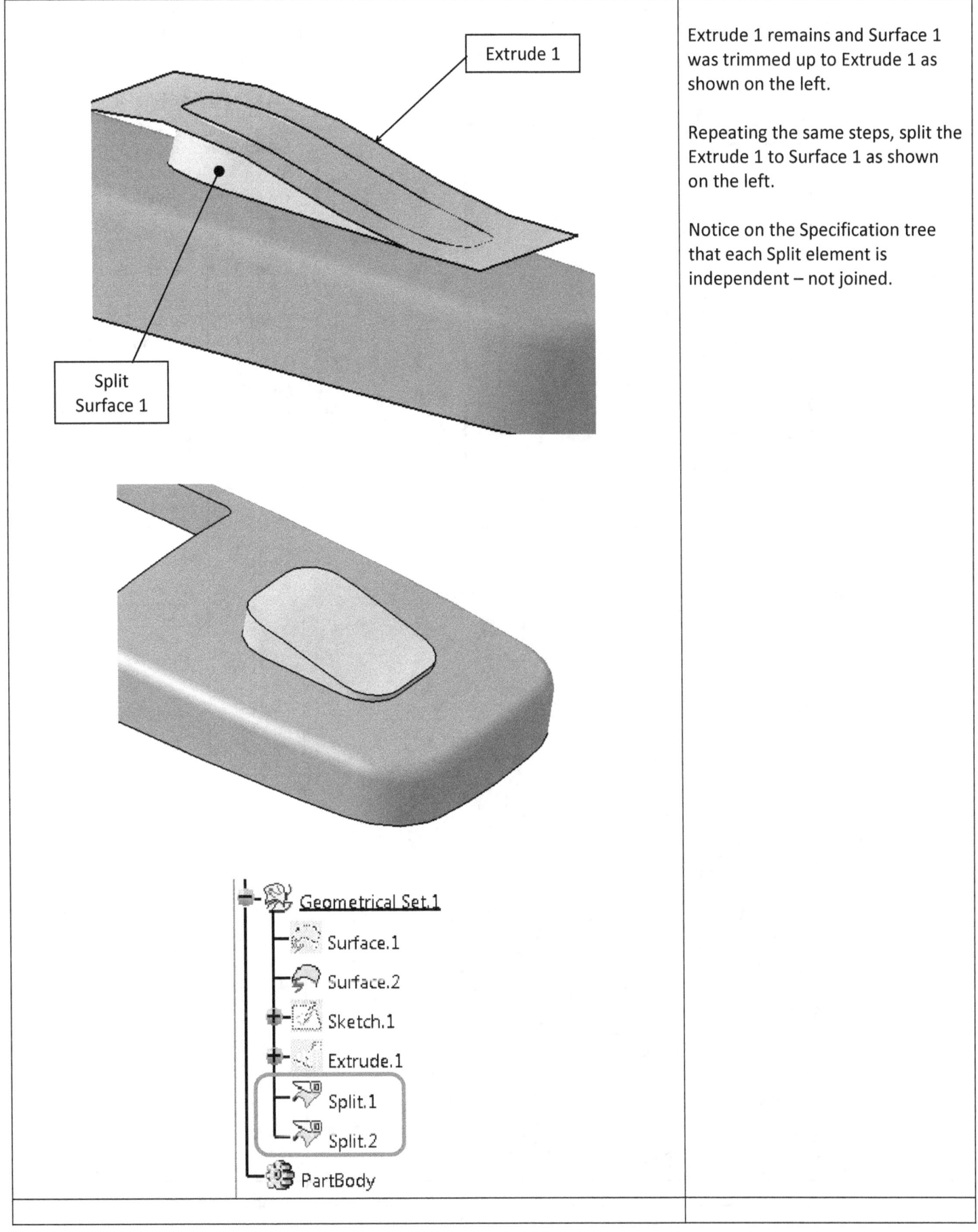

Extrude 1 remains and Surface 1 was trimmed up to Extrude 1 as shown on the left.

Repeating the same steps, split the Extrude 1 to Surface 1 as shown on the left.

Notice on the Specification tree that each Split element is independent – not joined.

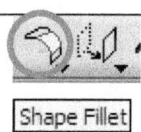

Shape Fillet

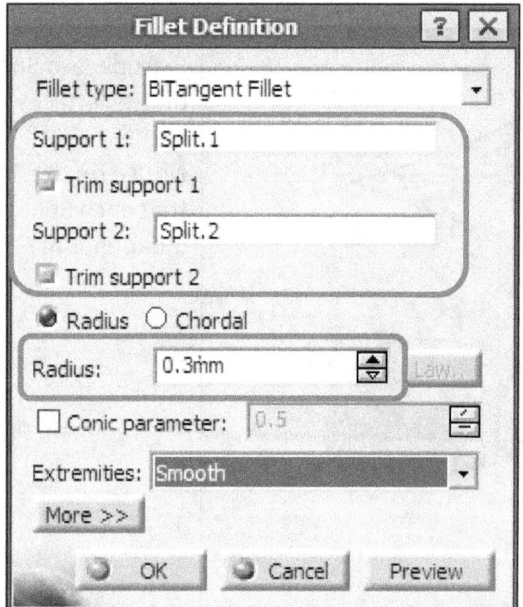

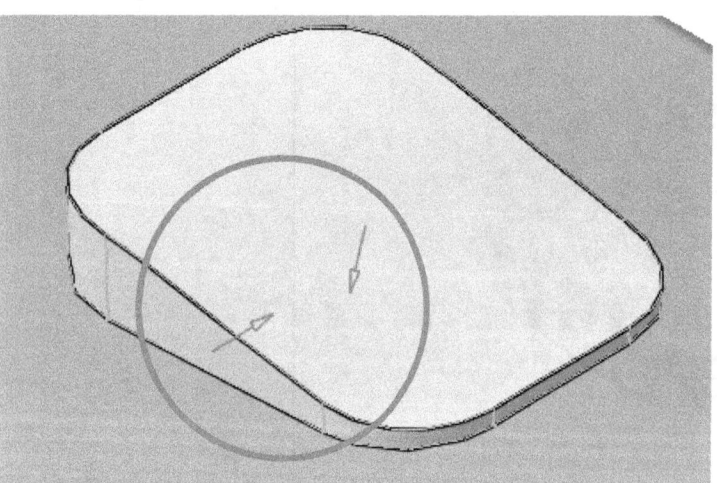

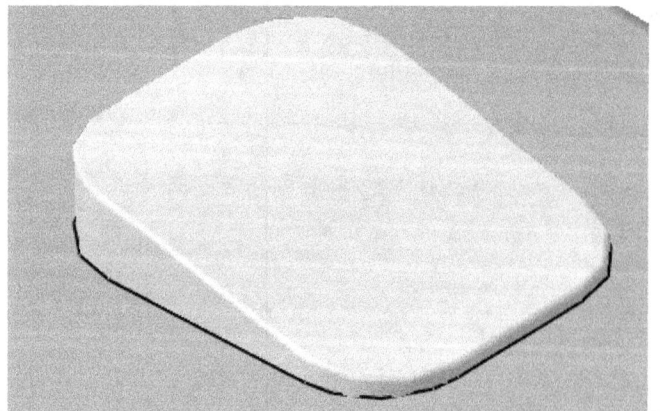

Now, add fillets around the button.

Click the Shape Fillet icon .

Select Split 1 for Support 1, and Split 2 for Support 2 in the Fillet Definition box as shown on the left.

Enter 0.3 mm as the Radius.

Ensure the arrow directions are facing inward as shown on the left.

Click OK.

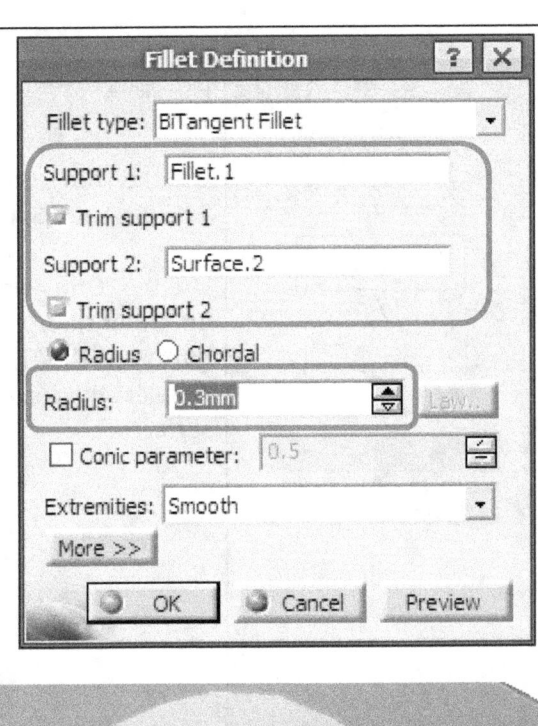

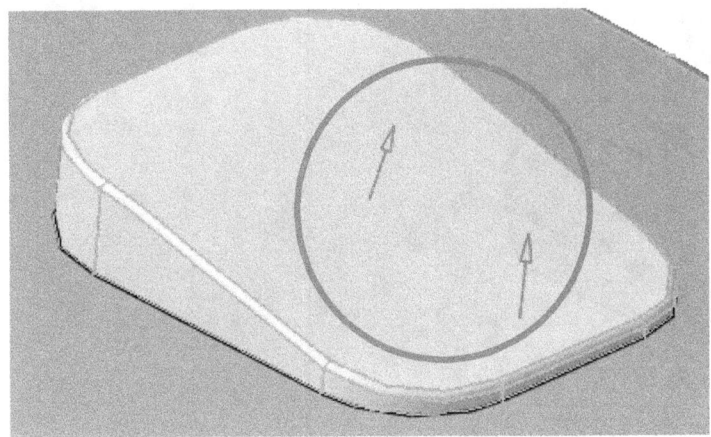

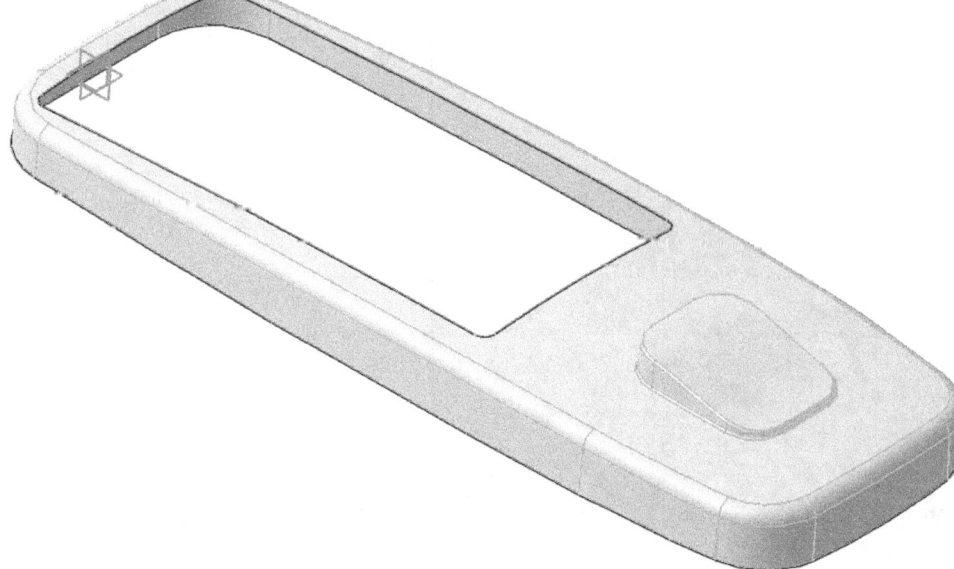

Add fillets between the body and the button.

Click the Shape Fillet icon .

Select Fillet 1 from the Specification tree for Support 1, and Surface 2 for Support 2 as shown on the left.

Enter 0.3 mm for Radius.

Ensure both arrows are pointing up.

Click OK.

Everything should now be connected.

Save this as "Split."

Open Trimming CATPart.

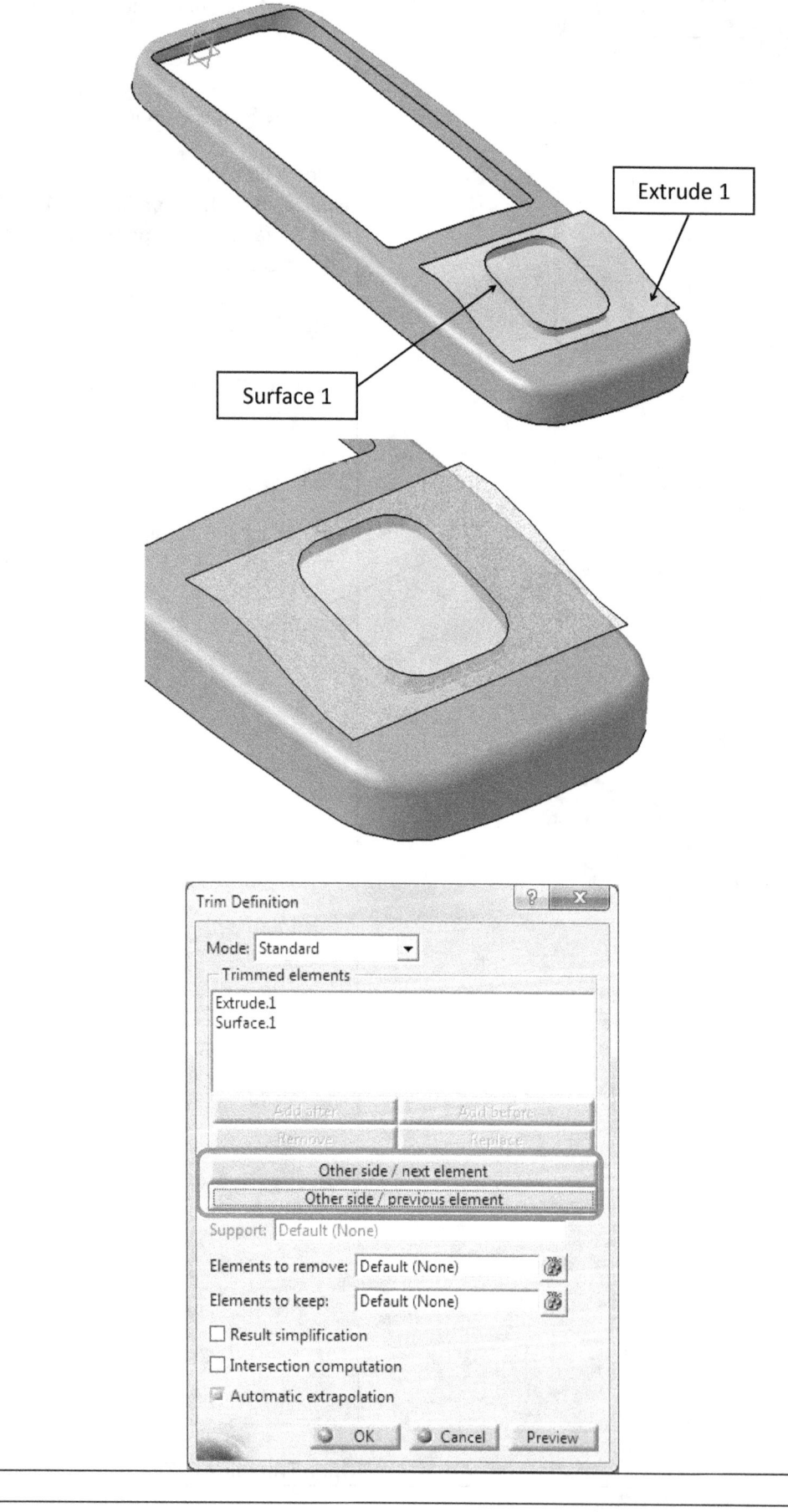

Set Units to mm.

Use the Trim icon , and select Extrude 1 and Surface 1.

Use the "Other side/next element" or "Other side/previous element" buttons in the Trim Definition box to select the correct sides of the surfaces.

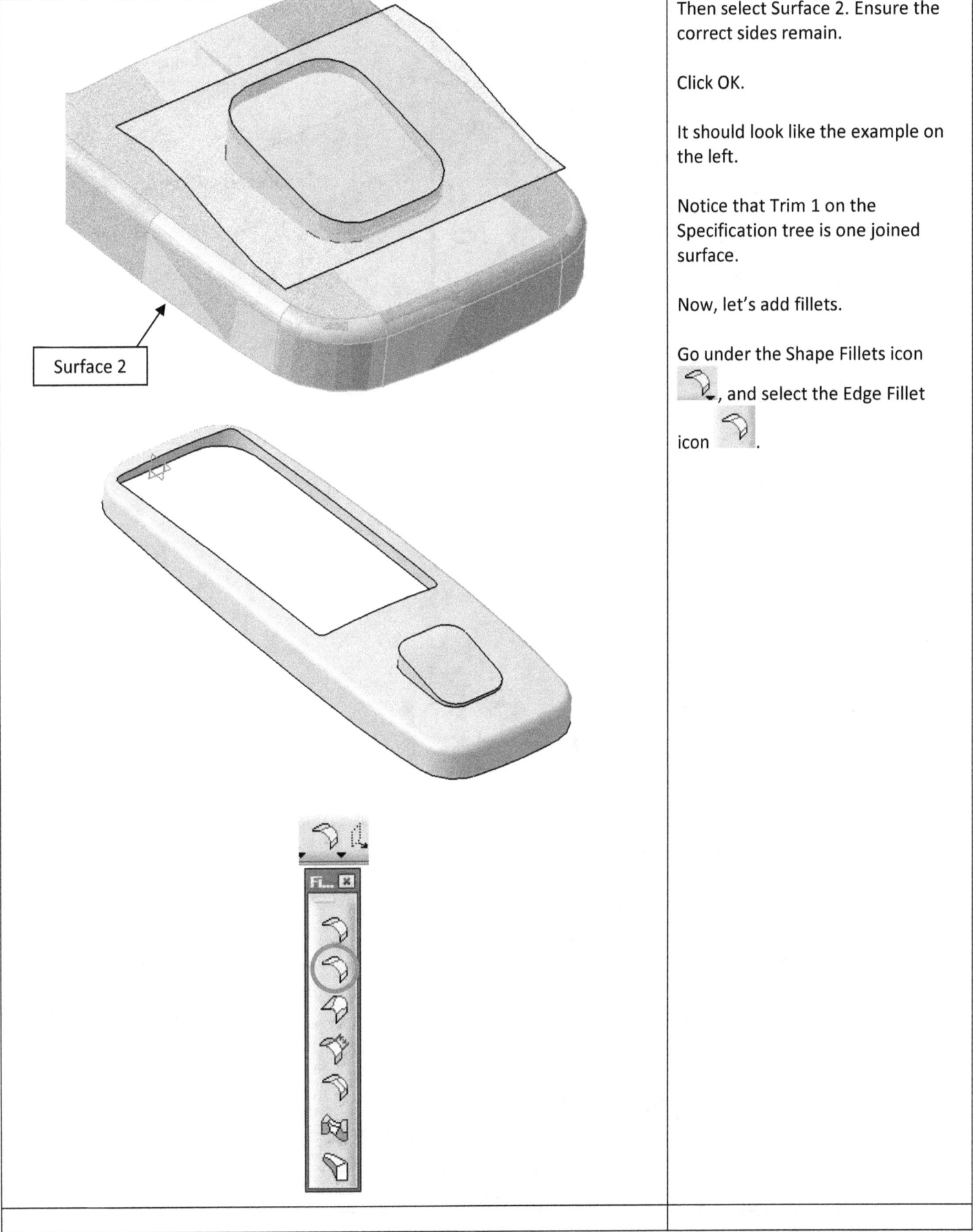

Then select Surface 2. Ensure the correct sides remain.

Click OK.

It should look like the example on the left.

Notice that Trim 1 on the Specification tree is one joined surface.

Now, let's add fillets.

Go under the Shape Fillets icon , and select the Edge Fillet icon .

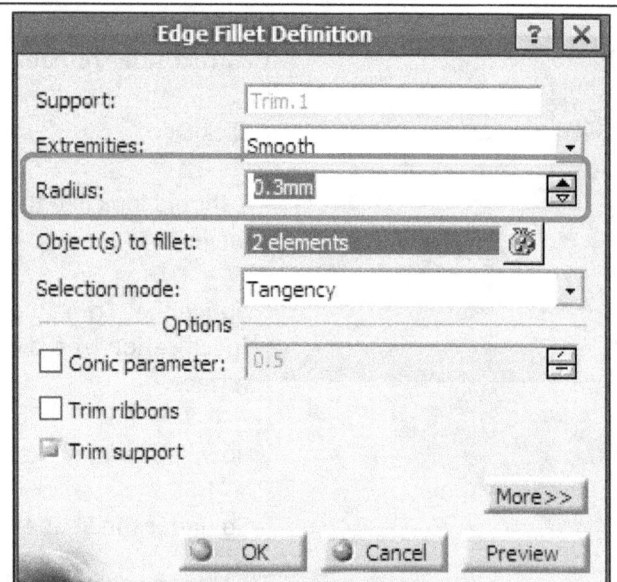

Enter 0.3 mm for the Radius.

Select both edges of Trim 1 as shown on the left.

Click OK.

Save this as "Trim."

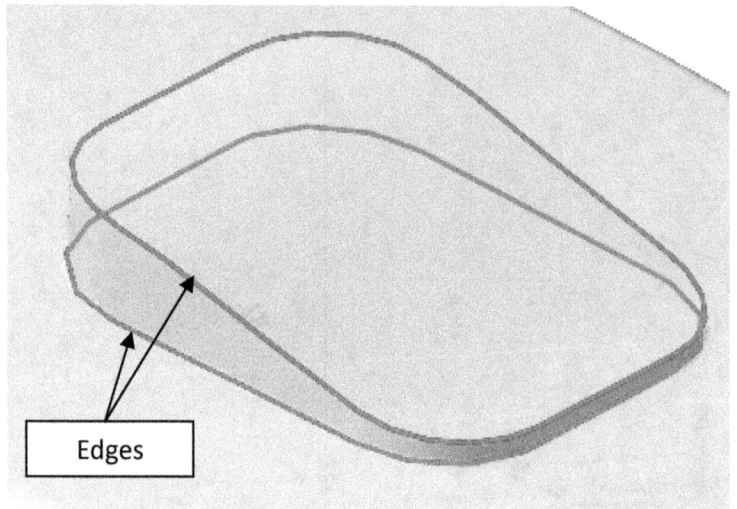

Edges

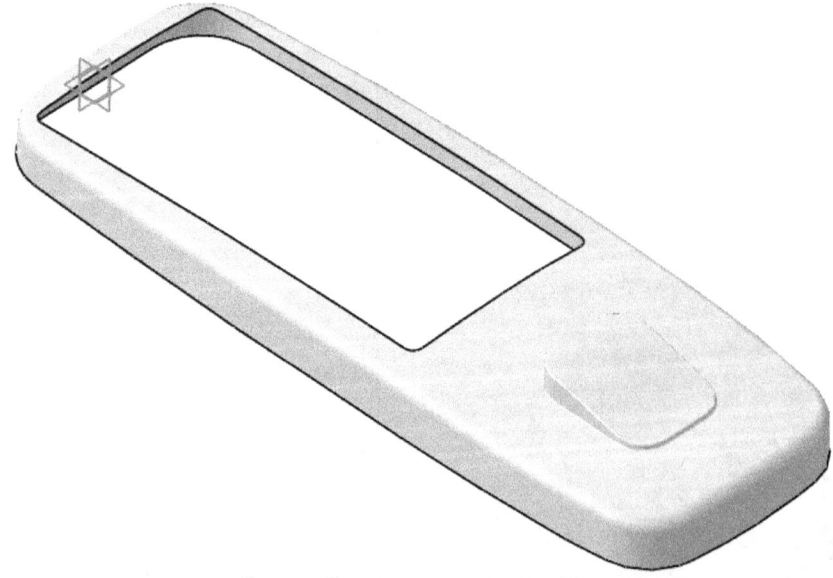

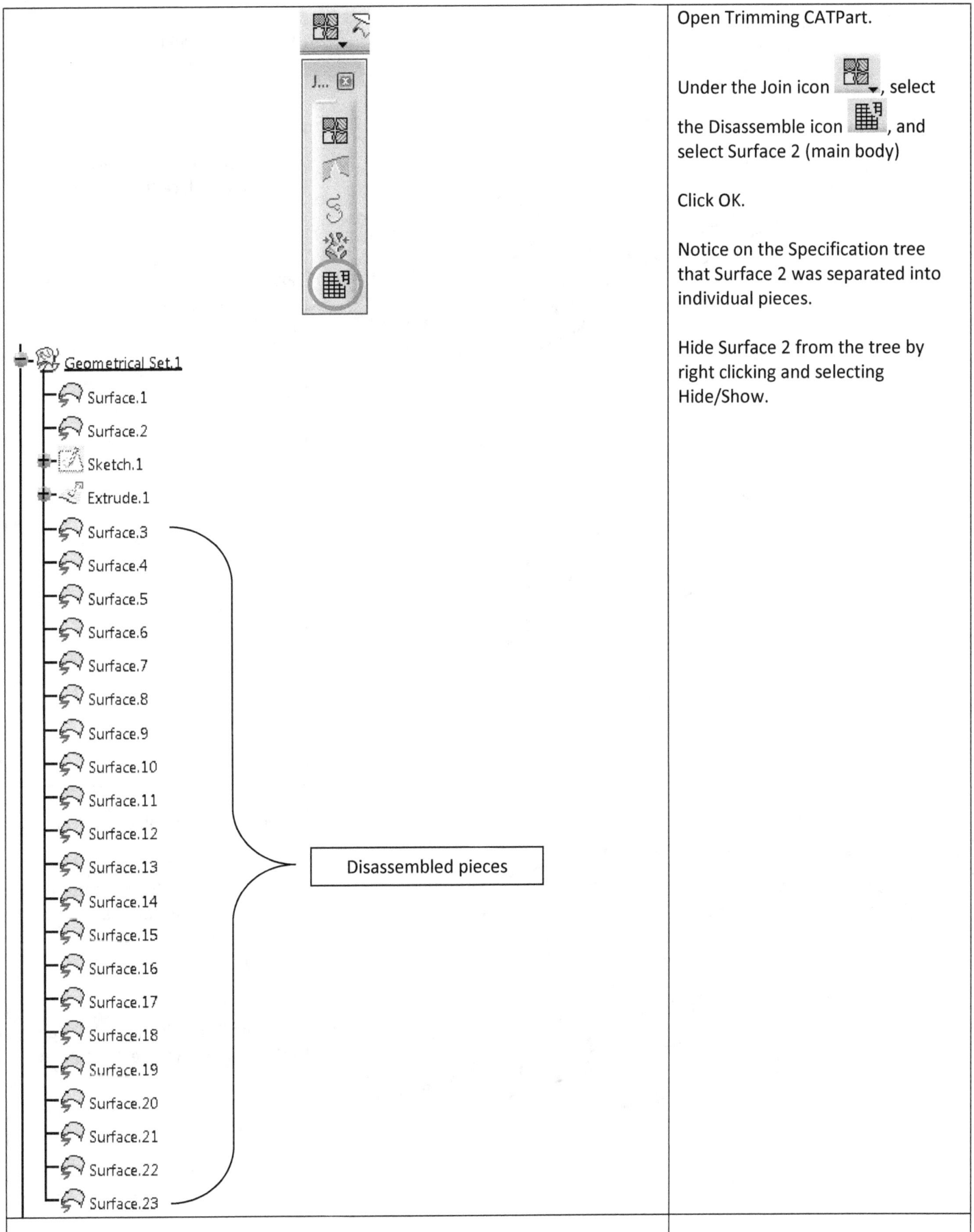

Open Trimming CATPart.

Under the Join icon, select the Disassemble icon, and select Surface 2 (main body)

Click OK.

Notice on the Specification tree that Surface 2 was separated into individual pieces.

Hide Surface 2 from the tree by right clicking and selecting Hide/Show.

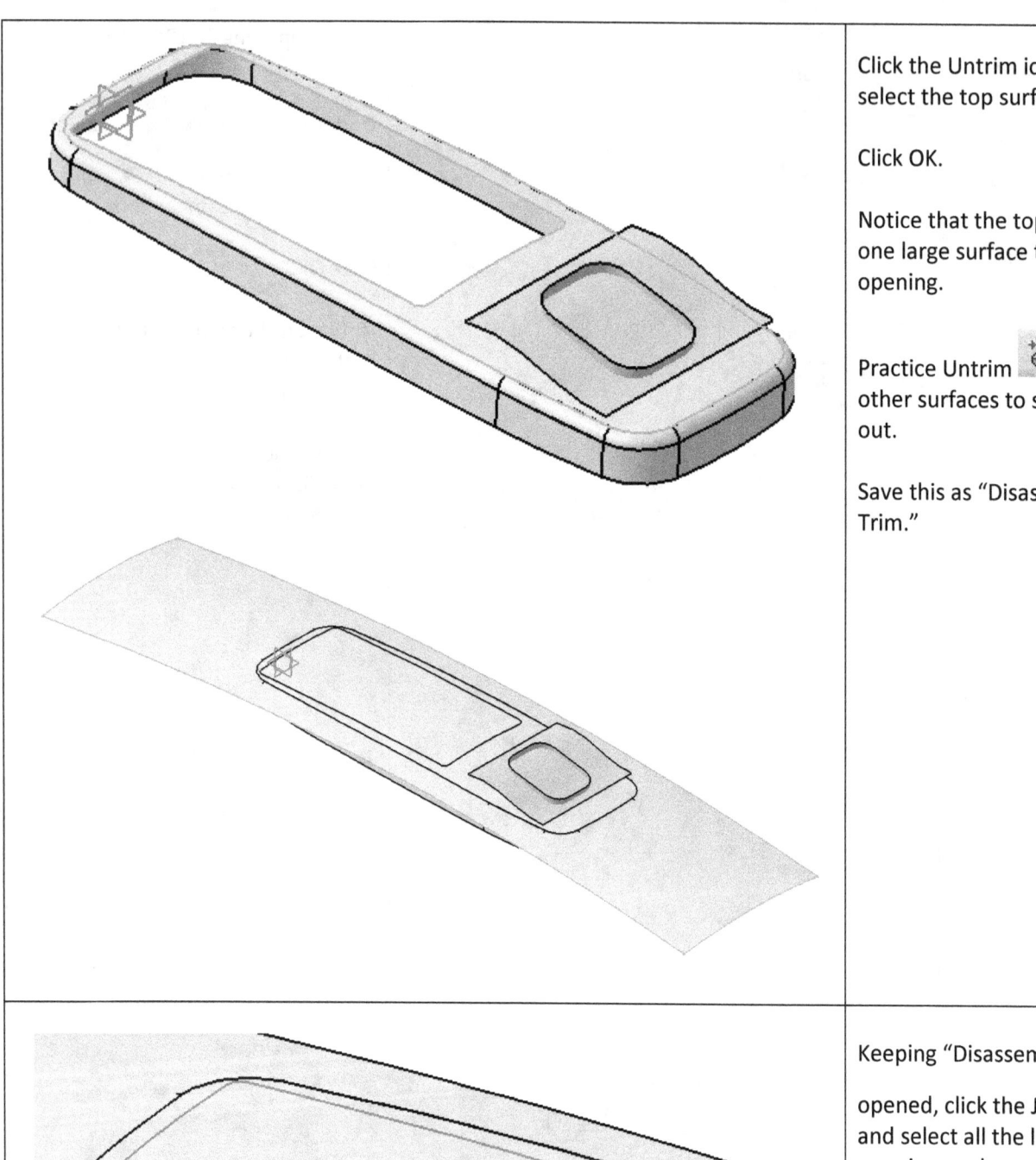

Click the Untrim icon , and select the top surface.

Click OK.

Notice that the top surface became one large surface that covers the opening.

Practice Untrim with some other surfaces to see how they turn out.

Save this as "Disassemble and Trim."

Keeping "Disassemble and Trim" opened, click the Join icon , and select all the lines of the opening as shown on the left.

Click OK.

Click the Projection icon , and select the Join you've just created.

Then select the untrimmed large top surface.

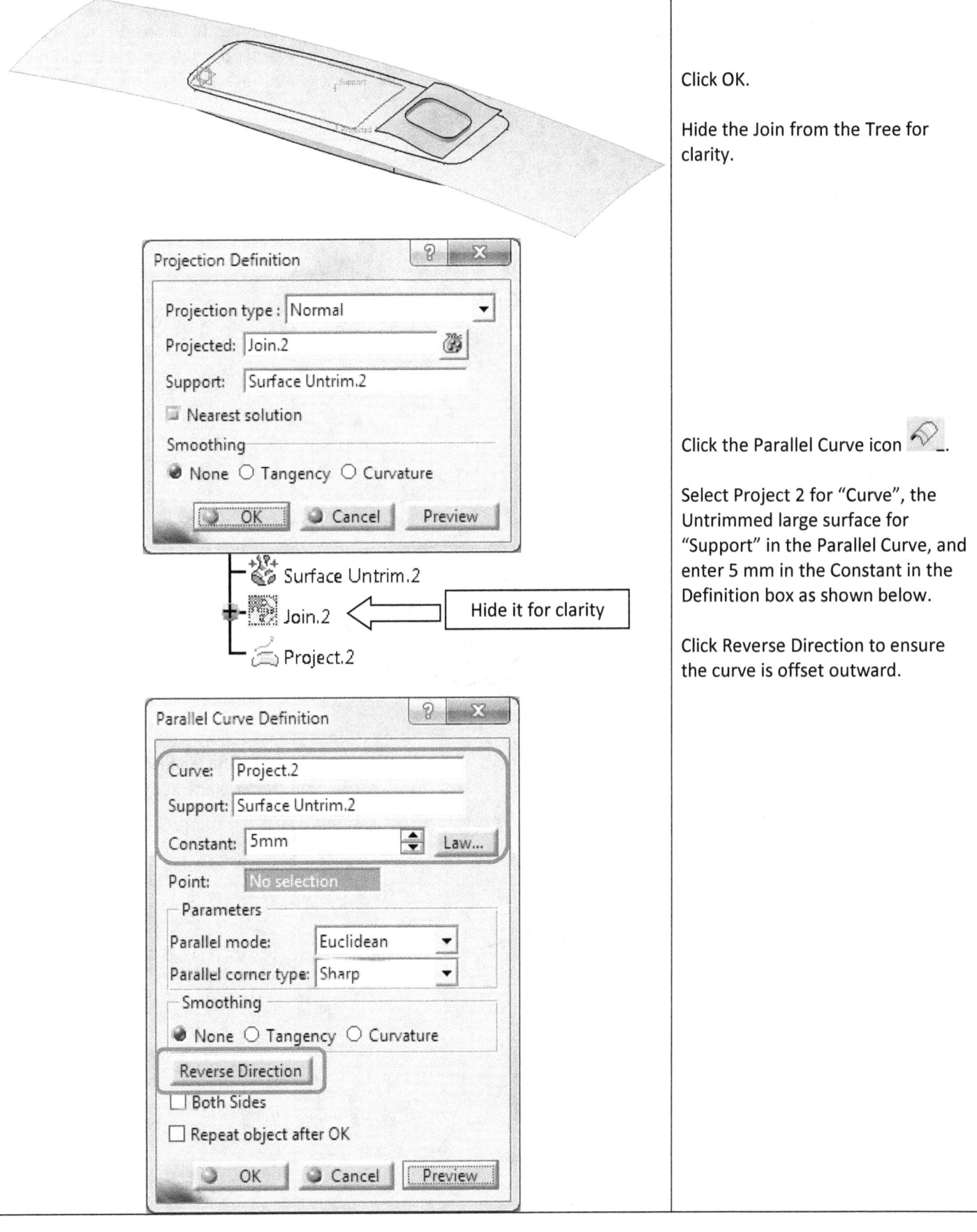

Click OK.

Hide the Join from the Tree for clarity.

Click the Parallel Curve icon.

Select Project 2 for "Curve", the Untrimmed large surface for "Support" in the Parallel Curve, and enter 5 mm in the Constant in the Definition box as shown below.

Click Reverse Direction to ensure the curve is offset outward.

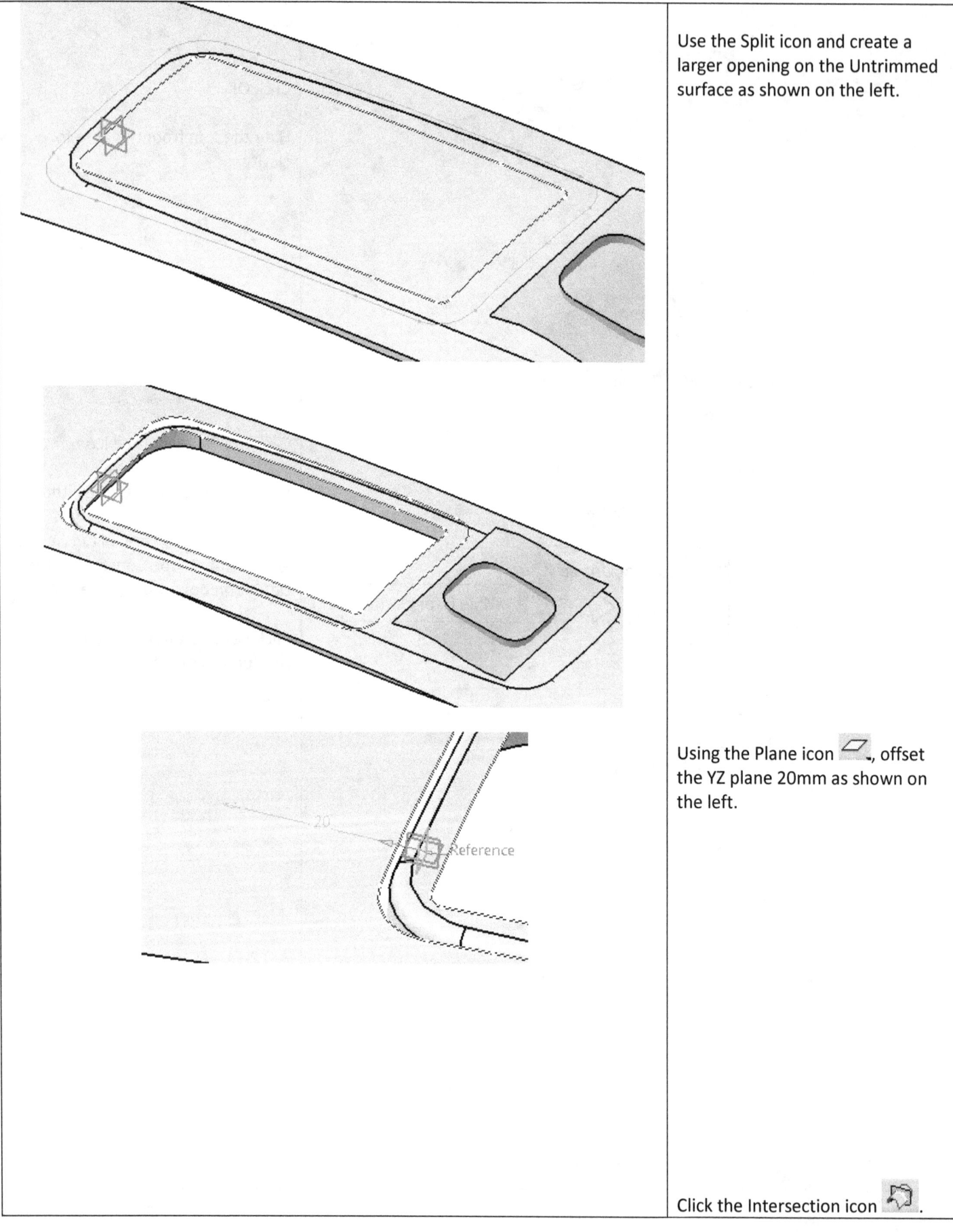

Use the Split icon and create a larger opening on the Untrimmed surface as shown on the left.

Using the Plane icon, offset the YZ plane 20mm as shown on the left.

Click the Intersection icon.

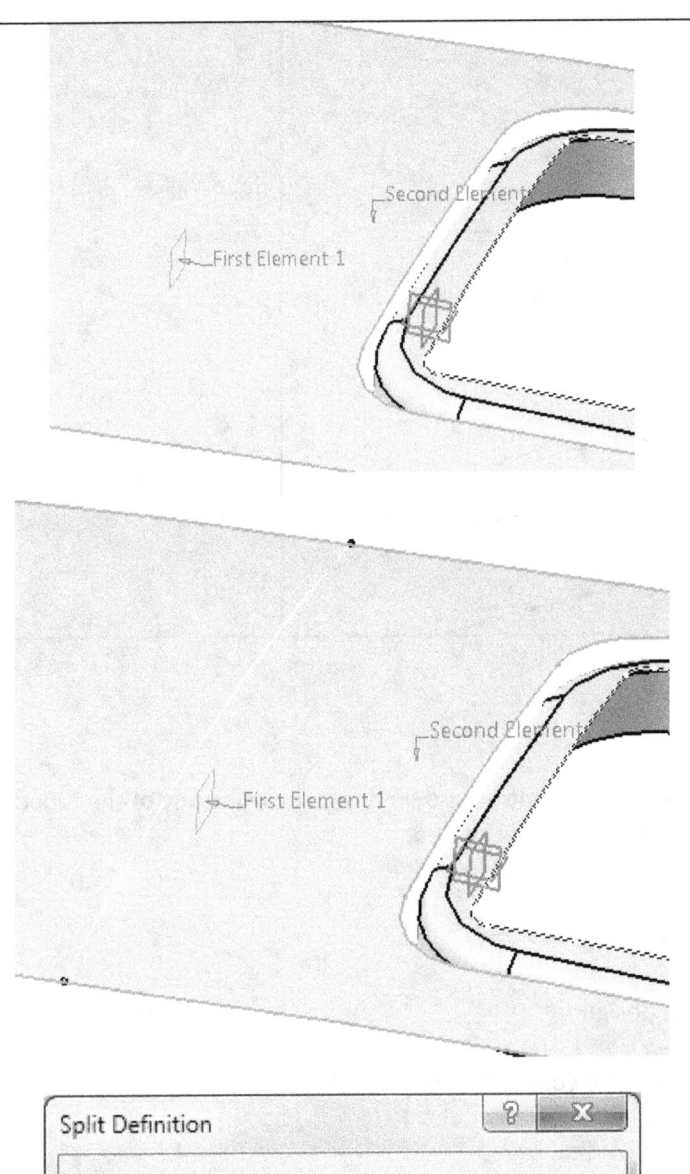

Select the new plane you just created and the large top surface that has a new larger opening.

Click Preview to see where those two elements intersect.

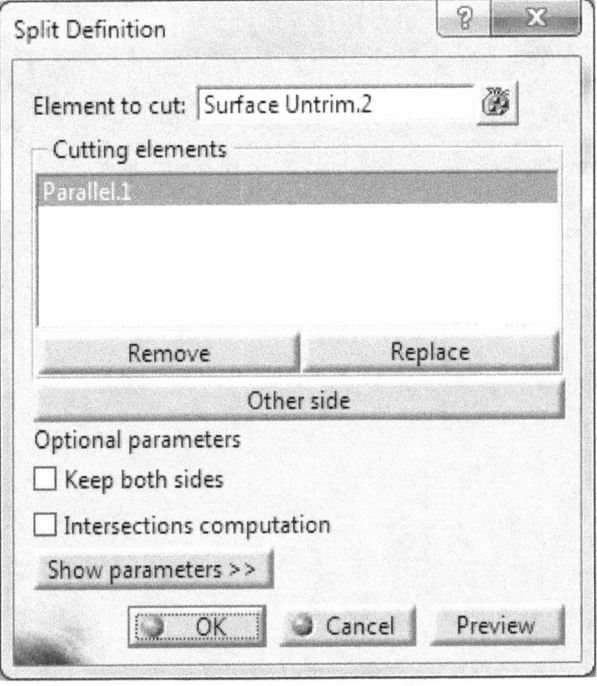

Split the top surface up to the intersection you just created as shown on the left.

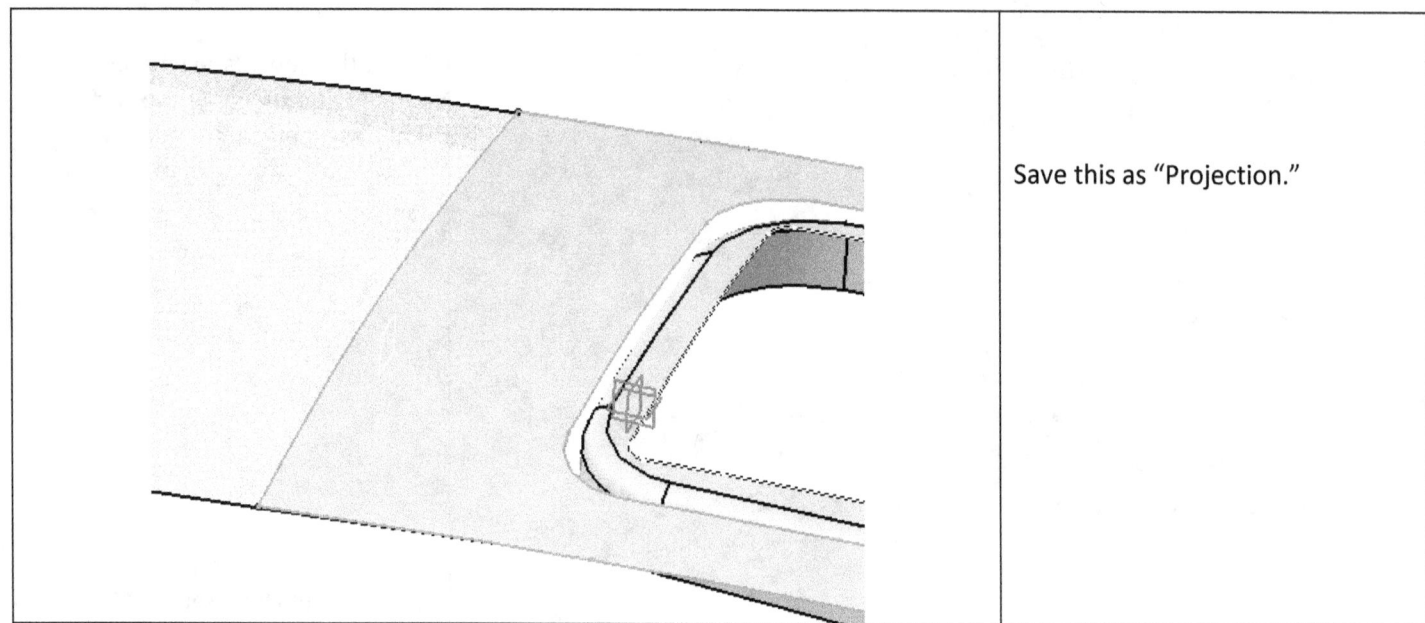

Save this as "Projection."

Chapter 8 Assignment

Based on what has been covered so far, alter this part to become your own design by using any of the functions in the Generative Shape Design Workbench. No solid modeling will be accepted.

Requirements:
- Change the button to be your own unique design
- Change the opening to be your own unique design
- Change the size and shape/form of the body to be your own unique design

www.ingramcontent.com/pod-product-compliance
Lightning Source LLC
Chambersburg PA
CBHW080819170526
45158CB00009B/2474